S

Is Animal Experimentation Ethical?

Bonnie Szumski and Jill Karson

INCONTROVERSY

ReferencePoint
Press®

San Diego, CA

ReferencePoint Press®

J
179.4
Szu

© 2012 ReferencePoint Press, Inc.
Printed in the United States

For more information, contact:
ReferencePoint Press, Inc.
PO Box 27779
San Diego, CA 92198
www.ReferencePointPress.com

LIBRARY OF CONGRESS CATALOGING-IN-PUBLICATION DATA

Szumski, Bonnie, 1958–; Karson, Jill
 Is animal experimentation ethical? / by Bonnie Szumski and Jill Karson.
   p. cm. — (In controversy)
 Includes bibliographical references and index.
 ISBN-13: 978-1-60152-174-3 (hardback)
 ISBN-10: 1-60152-174-X (hardback)
 1. Animal experimentation—Juvenile literature. 2. Animal experimentation—Moral and ethical aspects. I. Title.
 HV4915.S98 2012
 179'.4—dc22
                                                                    2011012464

Contents

Foreword 4

Introduction
Species and the Role of Animal
Experimentation 6

Chapter One
What Are the Origins of the Animal
Experimentation Debate? 10

Chapter Two
Is Animal Research Necessary for
Medical Progress? 24

Chapter Three
Should Primates Be Used in Research? 41

Chapter Four
Animals in Education 55

Chapter Five
What Are the Alternatives to Animal
Experimentation? 68

Related Organizations and Websites 79

Additional Reading 84

Source Notes 86

Index 91

Picture Credits 95

About the Authors 96

Foreword

In 2008, as the U.S. economy and economies worldwide were falling into the worst recession since the Great Depression, most Americans had difficulty comprehending the complexity, magnitude, and scope of what was happening. As is often the case with a complex, controversial issue such as this historic global economic recession, looking at the problem as a whole can be overwhelming and often does not lead to understanding. One way to better comprehend such a large issue or event is to break it into smaller parts. The intricacies of global economic recession may be difficult to understand, but one can gain insight by instead beginning with an individual contributing factor such as the real estate market. When examined through a narrower lens, complex issues become clearer and easier to evaluate.

This is the idea behind ReferencePoint Press's *In Controversy* series. The series examines the complex, controversial issues of the day by breaking them into smaller pieces. Rather than looking at the stem cell research debate as a whole, a title would examine an important aspect of the debate such as *Is Stem Cell Research Necessary?* or *Is Embryonic Stem Cell Research Ethical?* By studying the central issues of the debate individually, researchers gain a more solid and focused understanding of the topic as a whole.

Each book in the series provides a clear, insightful discussion of the issues, integrating facts and a variety of contrasting opinions for a solid, balanced perspective. Personal accounts and direct quotes from academic and professional experts, advocacy groups, politicians, and others enhance the narrative. Sidebars add depth to the discussion by expanding on important ideas and events. For quick reference, a list of key facts concludes every chapter. Source notes, an annotated organizations list, bibliography, and index provide student researchers with additional tools for papers and class discussion.

The *In Controversy* series also challenges students to think critically about issues, to improve their problem-solving skills, and to sharpen their ability to form educated opinions. As President Barack Obama stated in a March 2009 speech, success in the twenty-first century will not be measurable merely by students' ability to "fill in a bubble on a test but whether they possess 21st century skills like problem-solving and critical thinking and entrepreneurship and creativity." Those who possess these skills will have a strong foundation for whatever lies ahead.

No one can know for certain what sort of world awaits today's students. What we can assume, however, is that those who are inquisitive about a wide range of issues; open-minded to divergent views; aware of bias and opinion; and able to reason, reflect, and reconsider will be best prepared for the future. As the international development organization Oxfam notes, "Today's young people will grow up to be the citizens of the future: but what that future holds for them is uncertain. We can be quite confident, however, that they will be faced with decisions about a wide range of issues on which people have differing, contradictory views. If they are to develop as global citizens all young people should have the opportunity to engage with these controversial issues."

In Controversy helps today's students better prepare for tomorrow. An understanding of the complex issues that drive our world and the ability to think critically about them are essential components of contributing, competing, and succeeding in the twenty-first century.

Species and the Role of Animal Experimentation

The arguments for and against animal experimentation are varied and evoke strong emotions. Those on the side of animal rights believe that the suffering of animals cannot be justified, even in the face of clear medical advantages to humans. Those on the side of experimentation argue that the ends clearly justify the means—human health demands the sacrifice of animals.

Underlying these arguments is the concept of speciesism, the idea that certain species, usually those of higher intelligence, are more valuable than others. The idea that humans are the most valuable species to preserve is common among those who support animal experimentation. Others argue that humans should have no special dominion over animals and that the lives of animals must also count. Famed chimp researcher Jane Goodall remarks on the difficulty of sympathizing with animals over one's own species:

> In most cases, people will choose to sacrifice any animal to save or improve the quality of human life. In other words, in a scenario of "them" or "us," humans will always prevail. And this is hardly surprising. No matter how much a woman may love dogs or chimps, she will choose to sacrifice a dog or a chimp if told that this will save her child. Evolution has programmed us to make choices that ensure our genes will be represented in future generations. We choose to favor our own children

over the children of other people or other creatures. This is why those fighting for animal rights by using ethical and philosophical arguments, although they have made progress in changing attitudes toward animals, can never hope to bring all animal experimentation to an end by using these arguments alone.[1]

Goodall's argument appeals because she points out that our decisions are based on our emotional, gut desire to sustain our own flesh and blood over other considerations, even if these other considerations may not lead to a better conclusion. Her point, that humans will forget about any other good except for their own priorities, is clearly what makes the case of animal experimentation so loaded. Since humans are the ones that use animals for their own ends—whether it is as pets, as workmates, or as lab experiments, they will clearly make the decision to choose themselves over the animals.

A scientist involved in virus research takes a blood sample from a rhesus macaque monkey. The debate over using animals in medical research evokes strong opinions and emotions.

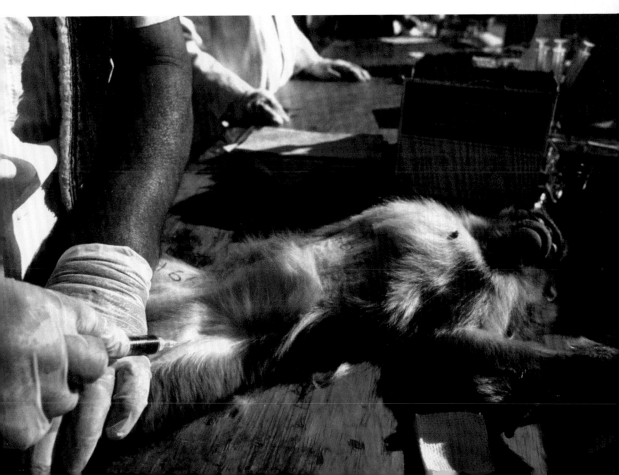

Yet even with this innate preference for their own species, do humans still have an obligation to animals? Should scientists preserve their dignity, minimize their suffering, and use animals only when it is absolutely necessary? Many scientists argue that this is indeed what happens. They claim that during the course of experiments, they name, care for, and begin to feel responsible for the animals they work with, even as they must distance themselves enough to inflict pain and suffering on them. In *When Species Meet*, Donna J. Haraway recounts the story of an old African man and the guinea pigs he cared for in a scientific experiment conducted in Zimbabwe in 1980. The experiment involved painting the guinea pigs with various poisons, allowing them to be attacked by tsetse flies, and watching whether the flies sickened or died. A young African girl, Nhamo, asks the old man, Baba Joseph, about the experiment:

> "It's cruel," agreed Baba Joseph, "but one day the things we learn will keep our cattle from dying." He stuck his own arm into a tsetse cage. Nhamo covered her mouth to keep from crying out. The flies settled all over the old man's skin and began swelling up. "I do this to learn what the guinea pigs are suffering," he explained. "It's wicked to cause pain, but if I share it, God may forgive me."[2]

Haraway uses this anecdote as a way of introducing the concept that humans are still obligated to the animals they must use for their own needs. She writes: "Using a model organism in an experiment is a common necessity in research. The necessity and the justifications, no matter how strong, do not obviate [remove] the obligations of care and sharing pain. How else could necessity and justice (justification) be evaluated in a mortal world in which acquiring knowledge is never innocent?"[3]

These concepts are behind many of the arguments found in the animal experimentation debate. Humans may continue to use animals for their own purposes, as Goodall writes. But she and many others argue that humans cannot, and should not, forget the shared obligation that one species owes another to have clear

"No matter how much a woman may love dogs or chimps, she will choose to sacrifice a dog or a chimp if told that this will save her child."[1]

— British primatologist Jane Goodall.

goals and minimize suffering. Researchers and others contend that people's shared relationships with animals, and their knowledge—mostly through pet ownership—of those animals' capabilities to suffer, to experience joy, and to devote themselves to people's interests, must inform the lab decisions.

In the end the African man cared about the guinea pigs, the cows, and the effect the research would have even on his soul. Perhaps it is only with this consciousness that humans can engage in the animal experimentation debate.

Facts

- A July 2009 survey by the Pew Charitable Trust found that 93 percent of scientists support animal testing, while only about half of the public supports it.

- The National Institutes of Health spends roughly $31 billion each year on medical research, much of which relies on animal models.

- Animals are used in experiments in three main areas: biomedical research, education, and toxicity testing of household substances.

What Are the Origins of the Animal Experimentation Debate?

The controversy over animal experimentation divides along a few fundamental principles that have informed the debate since its inception. The process of understanding human anatomy and physiology, improving human health through that understanding, and developing new medical procedures, vaccines, and other breakthroughs has always greatly relied on animal experimentation. The history of medicine is replete with new discoveries first being tried on animals. Medical research supporters argue that these experiments and the animal suffering that inevitably results are the price that must be paid for human understanding.

But it is suffering that is at the heart of the antiexperimentation debate. Philosopher Jeremy Bentham argued in 1789 that the question of whether to use animals was not "can they reason?" nor "can they talk?" but rather "can they suffer?"[4] It is this question that rallies those who oppose the use of animals in science. Because animals cannot reason but can know pain and suffering, can it be moral to allow their suffering? This question, more than any other, has informed the animal experimentation debate from its beginnings to the present.

Beginnings of Animal Protection

The animal experimentation controversy began in England in the late 1700s. There, people were trying to gain protection for domestic and farm animals to protect them from cruel treatment. The first organization to formally take on the issue of animal cruelty was the Society for the Prevention of Cruelty to Animals (SPCA), formed in 1824. The group not only defined what constituted cruelty, but brought charges against people who treated animals cruelly. The SPCA successfully prosecuted many violators of animal cruelty laws, which were passed in England to prevent mistreatment of farm animals. Because of their actions, Queen Victoria blessed their efforts with her patronage, and the SPCA became the Royal Society for the Prevention of Cruelty to Animals (RSPCA) in 1840.

The RSPCA became involved in animal experimentation in the 1870s when it sought to regulate conditions for animals used in scientific research. At that time, animals were used to teach medical students anatomy. Most experiments were done on live, unanesthetized animals that were immobilized by restraints. The group worked to pass the Cruelty to Animals Act in 1876, a law that required researchers to apply for an annual license, outlawed performing multiple operations on a single animal (which was normally not anesthetized), and allowed the government to mandate guidelines for particularly painful procedures. The act was highly criticized by researchers, who thought of it as a manifesto against science. These early efforts, then, clearly demarcated the philosophical division between science and antivivisectionists, those who oppose the use of animals in scientific research.

Animals Lead to Understanding of Principles

The work of French physiologist François Magendie and Scottish anatomist Charles Bell in 1822 further illustrates that divide. These men are credited with the discovery that the dorsal and ventral spinal nerve roots serve different functions in the central nervous system. They found that the dorsal roots conduct sensory information and the ventral roots conduct motor information. This information was crucial to understanding how the nerves interact with the spine. The entire basis for the field of neuroscience began with these basic principles. Magendie conducted his research using six-week-old puppies. He cut the dorsal and ventral roots from the spines of these puppies to observe loss of sensation and movement. Magendie repeated his studies on other animals.

Both Bell and Magendie performed these operations on unanesthetized animals. Although Bell began his research earlier that Magendie, his squeamishness over performing surgery on a live, unanesthetized rabbit made his research less defining than Magendie's. Bell ended up not being able to continue with the required experiments. In his private correspondence he wrote: "I cannot proceed without making some experiments, which are so unpleasant to make that I defer them. You may think me silly, but I cannot perfectly convince myself that I am authorized in nature, or religion, to do these cruelties."[5]

Thus, two researchers, working on the very same problem, were divided over the issue of animal suffering. At the time, many experiments relied on the animal being awake and responsive. The animal, though restrained, received little or no anesthesia. In Magendie's and Bell's experiments, the price of studying the science of anatomy came with the cost of live animals' agony.

In an example from around the same time, Claude Bernard researched curare, the toxin placed on darts by South American tribes to paralyze their prey. Bernard's research on the toxin, performed on animals, would eventually lead to an understanding of how certain drugs work in the human body. Bernard defended the use of animals in his experiments this way: "I think we have the right [to experiment on animals] wholly and absolutely. It would be strange indeed if we recognized man's right to make use of animals in every walk of life, for domestic service, for food, and then forbade him to make use of them for his own instruction in one of the sciences most useful to humanity."[6]

"I cannot proceed without making some experiments, which are so unpleasant to make that I defer them. You may think me silly, but I cannot perfectly convince myself that I am authorized in nature, or religion, to do these cruelties."[5]

— Scottish anatomist Charles Bell.

The Effect of Darwin's *Origin of Species*

Interestingly, it was a nineteenth-century scientific journey that furthered the cause of those who opposed animal experimentation. In *The Origin of Species* (1859) and *The Descent of Man* (1871), Charles Darwin argued that humanity evolved from lesser animals and was not initially born with the power of reason. These ideas, radical for their time, clearly showed that there was a link between humans and nonhumans. Darwin argued that although humans had evolved more in some areas than animals, animal biology and human biology were clearly similar to one another, and that animals experienced pain.

Darwin himself compromised on the issue of animal experimentation. He firmly believed in the necessity of animal experimentation, though he advocated for anesthetizing the animals first. In 1881 he wrote, "I know that physiology cannot possibly progress except by means of experiments on living animals, and I feel the deepest conviction that he who retards the progress of physiology commits a crime against mankind."[7]

Scientist Claude Bernard (center, with apron) conducts animal research in his laboratory in 1899. Bernard firmly believed in the necessity of animal experimentation. His work contributed to an understanding of how drugs work in the human body.

Animal Enterprise Terrorism Act

In February 2008 the husband of a biomedical researcher at the University of California–Santa Cruz was physically assaulted outside of his home. His attackers were animal rights activists who opposed his wife's use of animal models in her study of breast cancer. Another researcher affiliated with the same institution had his house firebombed. Although violent attacks like these are rare, attempts by animal rights groups to liberate animals from the laboratory have become increasingly common. These often take the form of undercover investigations of animal farms or research laboratories, in which investigators may break in or misrepresent themselves to gain admittance. They may surreptitiously film the animals and their living conditions and sometimes even release the animals from captivity.

To counter these types of activities, the US government enacted a series of tough laws called the Animal Enterprise Protection Act in 1992, which was amended to the Animal Enterprise Terrorism Act (AETA) in 2006. This legislation gives authorities the right to prosecute and convict individuals who threaten animal enterprises, which include commercial or academic organizations that use or sell animals or animal products for profit or food, agriculture, education, research, testing, and other activities. Under AETA, actions that damage property or cause injury, or the releasing of nonhuman animals, are criminal and carry stiff penalties.

Other Publications Spawn Concern over Animals

Darwin's musings spawned a spate of writings about animals as creatures capable of many human emotions and deeper feelings such as loyalty, complex reasoning, and emotional and physical suffering. The 1877 publication of the novel *Black Beauty*, Anna

Sewell's depiction of the life of a horse from the point of view of the horse, seemed to epitomize these ideas. The book's effect on society was profound; it became a best seller and exposed people to the notion that an animal could have complex feelings. The story was especially popular with children, whose natural love of animals is easily touched. *Black Beauty* became the first of many books that focused on animal cruelty, including the 1893 publication of *Beautiful Joe*, which told the story of an abused dog whose master cuts off its tail and ears as punishment. The dog eventually finds a loving home.

Although sympathy toward animals is a common sentiment today, animals were not kept as pets or thought of as companions in the 1800s. Animals were functional, and their owners required that they perform useful and profitable tasks. Though no doubt people became attached to their animals, it would have been considered a foolish luxury to keep a pet only as a companion. The ideas that animals could also have emotions and need protection against pain and suffering were not yet widely accepted notions, though treating animals kindly and well clearly benefited their owners by also keeping their value and performance high.

"*It would be strange indeed if we recognized man's right to make use of animals in every walk of life, for domestic service, for food, and then forbade him to make use of them for his own instruction in one of the sciences most useful to humanity.*"[6]

— French physiologist Claude Bernard.

The Movement Spreads to the United States

Though the SPCA began in Britain, its efforts led to counterparts in the United States, with some states, including New York and Massachusetts, passing anticruelty laws in the early 1800s. In the United States the antislavery movement during the 1860s provided a model for other social justice movements, including the early animal rights movement. Activist Henry Salt drew the parallel directly when he stated, "The present condition of . . . animals is [in] many ways analogous to that of the negro slaves . . . the same exclusion from the common pale of humanity; the same hypocritical fallacies to justify that exclusion; and as a consequence, the same deliberate stubborn denial of their social rights."[8]

Though the movement in the United States began with preventing cruelty to domestic and farm animals, as it had in Britain, it soon expanded to efforts aimed at preventing animal experimentation. Once the United States began to become more urban at the turn of the twentieth century, the idea of keeping pets caught on. Keeping small domestic animals for the pleasure of their company became common, and humanitarian efforts to protect them caught hold more quickly. The American branch of the SPCA (known as the ASPCA) formed animal shelters for unwanted animals and became a voice for other humanitarian efforts toward animals.

Antivivisection in America

The first American Anti-Vivisection Society (AAVS) was formed in 1883 by Caroline Earle White, who was also involved in the ASPCA. White formed the society after an American researcher, S. Weir Mitchell, approached her about purchasing dogs from the ASPCA shelter for use in experiments. White refused and realized that an American Anti-Vivisection Society would help her organize efforts to oppose the use of pets in research. White's group grew quickly, and it attacked the scientific community vigorously.

The AAVS claimed that scientific experiments on animals often had little or no application to human disease. Often animals reacted differently from humans to the same procedures, and thus animal lives were wasted for little or no scientific gain. The AAVS also argued that animal experimentation made researchers immune to human suffering. Hoping to trigger an emotional response from the general population, the group publicly exposed the suffering of animals in labs by publishing in newspapers photos and stories of dogs that had crushed spines, burned paws, and parts of their brains removed.

Yet researchers had a litany of medical advances to lend credence to their arguments, and they denied that animals suffered with the same intensity as humans. Major advances such as Joseph Lister's germ theory, Robert Koch's research on the causes of cholera, the treatment of syphilis, and the rabies vaccine all involved experimentation on animals. In some instances the AAVS was its own worst enemy, initially mocking the idea of vaccinations, for example, which clearly became an essential disease-fighting tool.

"I know that physiology cannot possibly progress except by means of experiments on living animals, and I feel the deepest conviction that he who retards the progress of physiology commits a crime against mankind."[7]

— English naturalist Charles Darwin.

In addition to countering AAVS attacks, the medical community announced in 1908 the creation of the Council on Defense of Medical Research. The council organized routine visits to investigate animal care in laboratories. Proposals the group promised to implement included consistent rules for how laboratory animals should be treated, guidelines for obtaining stray dogs from procurers, sanitation standards for kennels, a policy for when it was important to use animals, and humane euthanasia for all animals used in scientific experiments. In a way, such steps could be regarded as victories for the antivivisectionists, whose consistent attacks on the treatment of experimental animals forced the medical community to make concessions for public acceptance.

A Public Battle

The battle between antivivisectionists and the medical community played out in newspapers, magazines, and other public forums, as antivivisectionists were able to get major American magazines such as *Vogue* and *Harper's* to publish horrific stories of animal experimentation. They also courted celebrities who were against the practice. Mark Twain actively advocated for the antivivisectionist cause: "I am not interested to know whether vivisection produces results that are profitable to the human race. . . . The pains which it inflicts upon unconsenting animals is the basis of my enmity towards it."[9] In a Christmas story called "A Dog's Tale" that he wrote for *Harper's Monthly Magazine* in 1903, Twain follows the life of a domestic dog who is at first beaten by her owner, then blinded by a medical researcher who is lauded for his research. Such emotional attacks against animal experimentation resonated with the public.

War and Medical Progress

Between World War I and World War II, the United States increased in stature as a world power. Its researchers also gained fame for medical breakthroughs. New medicines developed through the use of animal experimentation saved thousands, if not millions, worldwide. Insulin, vitamins, and sulfa drugs were just a few of the miracle medical treatments. As the United States industrialized, formal institutions and private companies began to specialize in medical research. The first, the Rockefeller Institute for Medical Research, was founded in 1901. It was followed by many others that were founded in the beginnings of the twentieth century, including Merck, Parke-Davis, and Eli Lilly. The government also lent a hand to research when it announced the creation of the National Institutes of Health in 1930.

The development of these laboratories effectively shielded the public from much of the animal experimentation. In addition, the public was awed by scientific progress, and few asked about the fate of the animals sacrificed to accomplish it. In one example the Russian dog Laika became the first living being to orbit the Earth in 1957. Though the dog died of overheating and stress shortly after takeoff, few questioned the use of the animal in the name of scientific progress. Although still active, the voice of the anti-vivisectionists was somewhat hampered by the belief in scientific progress.

The Modern Animal Rights Movement

The modern animal rights movement is dated to the publication of Australian philosopher Peter Singer's book *Animal Liberation* in 1975. Singer's book covered many areas of animal suffering, including slaughter in agriculture and scientific research. In it he argued that the use of animals in experiments should be subject to guidelines. For instance, scientists should only use animals in research after exhausting all other means.

His book spawned groups that would use tactics to gain publicity for the rights of animals, including People for the Ethical Treatment of Animals (PETA), the Animal Legal Defense Fund, and the Animal Liberation Front. These groups organized break-ins and seizures at animal laboratories to expose animal research in the 1970s and 1980s. Their efforts brought many painful experiments into the public eye, including the then-common practice of testing cosmetics on the eyes of rabbits. They also brought to light painful experiments on higher-level animals, such as primates, and the sometimes deplorable conditions such animals endured. Their efforts brought renewed interest in the regulation, though not the elimination, of animal experiments.

One of the most significant legislative acts to come out of the fight between these two camps was the passage of the federal Ani-

The Rise of Safety Testing

In addition to the millions of laboratory animals used each year in biomedical research, an unknown number of animals are used today to test the safety of consumer goods. In her book *For the Prevention of Cruelty: The History and Legacy of Animal Rights Activism in the United States*, Diane Beers describes the rise of product testing in the 1960s and 1970s:

> By the 1960s, nearly every product that landed in shop or showroom was first tested on the planet's nonhumans. Government regulations required toxicity testing on animals for only some items (pharmaceuticals and food, for example), but they recommended testing more generally as an acceptable way to enhance product safety. In response, manufacturers drizzled everything from floor wax to laundry detergent into the acutely sensitive eyes of albino rabbits and force-fed everything from lipstick to talcum powder and glue to mice or rats. Corporate testing soon spread to another giant industry, automobile manufacturing. . . . Ford experimenters strapped conscious, unanesthetized baboons into cars or "impact sleds" and sent them crashing into barriers. Researchers frowned on the use of anesthesia because it interfered with the proper simulation of a human's crash responses.

Diane Beers, *For the Prevention of Cruelty: The History and Legacy of Animal Rights Activism in the United States*. Athens, OH: Swallow, 2006, p. 178.

mal Welfare Act (AWA) of 1966. At this time, antivivisectionists focused their energies on two main principles: the alleviation of suffering and the improvement of conditions for laboratory animals. Because of pressure on the medical community by animal advocates, a host of laws and procedures govern the use of animals today. Researchers who wish to use animals in their experiments must conform to a series of rigorous guidelines on how such animals will be housed and how their pain will be diminished, as well as prove that the research cannot be completed by any means other than the use of animals. The tension between the medical community and the antivivisectionists, then, brought necessary attention to the plight of animals while also acknowledging the need for such animals in the search for new medical developments.

Protections for Research Animals

Today the US Department of Agriculture (USDA) enforces the AWA, which regulates the care of many warm-blooded vertebrates, including rabbits, hamsters, guinea pigs, primates, cats, dogs, and other animals. The act does not monitor all animals, however; in 2002 a fifth amendment to the AWA excluded protection for certain animals, including mice and rats, the most common species of laboratory animals.

The AWA requires that all universities and other research laboratories using protected species establish and maintain an Institutional Animal Care and Use Committee (IACUC), which is responsible for ensuring compliance with AWA provisions. The IACUC is made up of at least 5 members, including veterinarians, animal care technicians, and other individuals; this group must approve all proposed research projects and inspect the facilities that house the animals. In addition, the IACUC oversees pain management—that is, the amount of pain and distress the captive animals will experience as part of the research and how it is treated. For example, of the 1 million animals used in research in 2007, over 600,000 were involved in experiments that caused pain. Of these, over 550,000 animals had their pain alleviated with drugs or other measures. Many hope that the existence of IACUCs will generate a strong commitment within the research community to minimizing animal discomfort, distress, and pain.

In addition, the US Public Health Service Policy on Humane Care and Use of Laboratory Animals oversees any institution that conducts animal research with federal funds. Specifically, the

"I am not interested to know whether vivisection produces results that are profitable to the human race. . . . The pains which it inflicts upon unconsenting animals is the basis of my enmity towards it."[9]

— American author Mark Twain.

policy requires that each institution have an IACUC as part of a comprehensive animal care and welfare program and that all institutions base their animal care on the *Guide for the Care and Use of Laboratory Animals.* This handbook, published by the National Academy of Sciences, provides instructions on how to care for laboratory animals in an ethical way. One portion of the guide, for example, advises researchers on the most humane way to euthanize laboratory animals. The Public Health Service policy, moreover, is broader than the AWA because it includes protections for rats, mice, birds, and even fish and reptiles.

Unaffiliated with the federal government, the American Association for the Accreditation of Laboratory Animal Care is an independent organization that promotes the welfare of research animals. Institutions join the association, which regularly inspects their research facilities and offers accreditation for those that maintain high standards of animal care. While it is not possible for these groups to police every activity at the many research laboratories across the country, many believe that these types of animal care measures will help generate greater sensitivity to the plight of laboratory animals and ensure that the vast majority of scientists treat their research animals humanely.

A technician in Russia prepares Laika for her journey into space. In 1957 the dog became the first living being to orbit Earth. She died from overheating and stress shortly after takeoff.

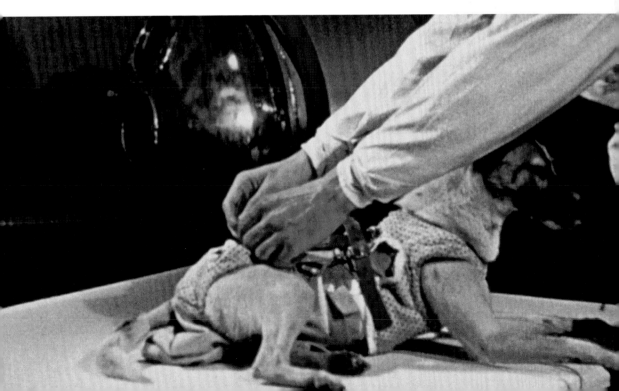

On the other hand, many animal rights activists charge that the AWA and other protections are hardly all encompassing and in fact do little to protect laboratory animals in captivity. As one animal rights group puts it:

> The AWA places no real restrictions on what can be done to an animal during an experiment. Animals are routinely subjected to addictive drugs, electric shock, food and water deprivation, isolation, severe confinement, caustic chemicals, burning, binding, chemical and biological weapons, radiation, etc. The "scientist" in question only has to say that a specific procedure is "necessary" for the experiment, and it is allowed. The goal is not to protect the animal; the goal is to insure that the experiment proceeds—at any cost.[10]

Animal Use Trends

It is difficult to assess exactly how many animals are employed in research laboratories each year. As part of its animal management, the USDA tracks these trends. By some accounts, it appears that fewer animals are used in research than when the department first started reporting statistics in 1973. According to the most recent reports, the number of research animals dropped from 1.6 million in 1973 to just over 1 million in 2007. This figure includes approximately 72,000 dogs, 22,000 cats, 70,000 primates, and 236,000 rabbits.

It is likely, however, that research facilities are simply replacing reportable animals—dogs, cats, primates, and other protected species—with mice and rats. Although these are not reported, they account for the vast majority of animals used in research because of their small size and because researchers can purchase them cheaply. Larry Carbone explains why these small animals are likely to remain the research animal of choice for many years to come:

> Mice, rats, and other small animals . . . have "job security" for several reasons. First, they are easily genetically modified, either by removing one or more of their native genes . . . or by inserting one or more genes from humans, other rodents, or even fireflies and jellyfish. Second, as with computers, technological advances have allowed the miniatur-

ization of many animal research projects. . . . Third, rodents and fish lack the political and public concern that larger mammals have; their exclusion from Animal Welfare Act coverage is itself sufficient to push many research institutions to deal exclusively in rodents and "lower" animals.[11]

Although many questions remain about the ethics of using animals in the laboratory, research using animal models, especially small rodents, will likely play a prominent role in modern medical research for years to come.

Facts

- Each year, many more invertebrates than vertebrates are used in animal research, yet experiments on invertebrates—including fruit flies, worms, and many other animals—are largely unregulated.

- According to the *Guide for the Care and Use of Laboratory Animals*, the text that governs animal welfare regulation in the United States, the use of anesthetics and analgesics to alleviate pain in research animals is an ethical imperative.

- A 2010 study at McGill University and the University of British Columbia discovered that mice, one of the most common research subjects, showed discomfort through facial grimaces when subjected to moderate pain stimuli.

- More than 90 percent of the animals used in research are not covered by the Animal Welfare Act that governs the treatment of animals. Birds, rats, and mice, for example, are excluded from the act's protections.

- Many research animals today are used to study behavioral phenomena. Worms, mice, rats, and birds are among the animals widely used to study social behavior and how animals interact with their surroundings.

Is Animal Research Necessary for Medical Progress?

In November 2006 Peter Singer, largely recognized as the founder of the modern animal rights movement for his seminal work *Animal Liberation*, was asked about his opinion on Parkinson's disease research that had been conducted on primates. The research, in which about 100 monkeys were used, identified a certain part of the brain as being involved in the disease. The researcher, Tipu Aziz, an Oxford neurosurgeon, estimated that the lives of approximately 40,000 Parkinson's sufferers had been improved by this research.

Singer, who spoke directly with Aziz in a BBC documentary, replied: "I think if you put a case like that, clearly I would have to agree that was a justifiable experiment. . . . I could see that as justifiable research."[12]

After the release of the documentary, subsequent discussions consumed pro- and anti-animal-experimentation websites. The former lauded that even a "radical" animal rights figure such as Singer had to acknowledge that animal experimentation had a place in society, while anti-animal-experimentation websites proclaimed that Singer never was a representative of animal rights and clearly was morally and ethically bereft.

Developing a Criteria for the Use of Animals

This relatively recent incident is at the heart of what may be the most important question of animal experimentation—what criteria justify the use of animals in research? This question and the discussion over what criteria should guide animal experimentation are not new. In 1831 British physiologist Marshall Hall proposed five principles that should guide research. As quoted in a Johns Hopkins University newsletter, these were:

> First, an experiment should never be performed if the necessary information could be obtained by observations; second, no experiment should be performed without a clearly defined and obtainable, objective; third, scientists should be well-informed about the work of their predecessors and peers in order to avoid unnecessary repetition of an experiment; fourth, justifiable experiments should be carried out with the least possible infliction of suffering (often through the use of lower, less sentient animals); and finally, every

A monkey is tested for the effects of marijuana on the eye. The criteria for experimentation on animals vary depending on the type of animal and the reason for the research. Criteria for experimentation on primates are more stringent than for other animals.

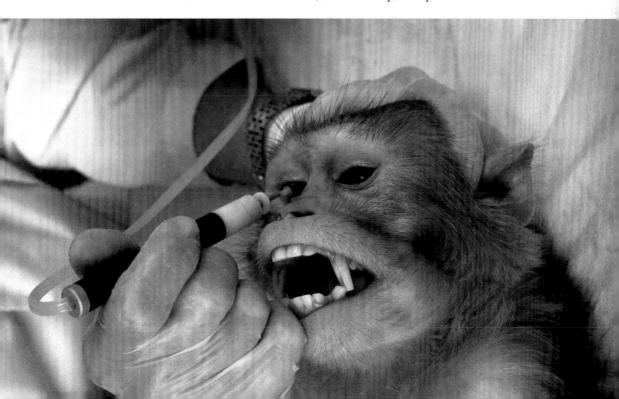

experiment should be performed under circumstances that would provide the clearest possible results, thereby diminishing the need for repetition of experiments.[13]

Many researchers continue to use these same principles to guide their research. Today these guidelines are encompassed in the "three Rs"—reducing the number of animals used, refining procedures to minimize pain and enhance animal well-being, and using replacement alternatives whenever possible. Many advocates say that using such criteria ensures that the lives of animals are safeguarded and that clear proof of human benefits will be maintained in order for an animal life to be appropriately sacrificed. Even though millions of animals, mostly rats and mice, are used annually, these animals provide necessary and crucial information that informs both human and animal health.

A Medical Fantasy

Advocates of animal experimentation credit it with recent new treatments for heart failure, Alzheimer's disease, Parkinson's disease, Lou Gehrig's disease (also called ALS), obesity, sleep disturbances, diabetes, genetic diseases, cancer, and emphysema. Carl Cohen, a professor of philosophy at the Residential College of the University of Michigan in Ann Arbor, makes the argument that animal experimentation is crucial:

> Animals can*not* be adequately replaced. Substituting non-animal methods for testing in . . . most medical research is a wishful fantasy . . . in the vast majority of biomedical investigations, there are absolutely no satisfactory replacements for animal subjects, and there are none even on the horizon, *because the safety and efficacy of an experimental drug or procedure must be determined by assessing its impact on a whole, living organism.*[14]

Cohen uses the polio vaccine as just one of many examples of the necessity of animal experimentation: "How many have been spared misery and death by this one great step in medical science we can hardly guess. But about this wonderful vaccine and its successors we do know one thing for certain: *It could not have been achieved without the use of laboratory animals.*"[15]

Cohen and other supporters argue that animals cannot be replaced. Replicating diseases and procedures in animals and testing drugs on animals, for example, are essential steps in developing new treatments because these studies make such things safe for humans. When such experiments fail and animals die in the process, it is still a much less tragic fate than if such procedures were tested on humans first and humans died. Many advocates make the argument that the alternative is far worse—when pitted against experimentation in people, animal experimentation for safety and efficacy is the only way.

Designer Tumors

One example advocates cite as to whether research is necessary is a new treatment for people with head and neck cancer. In this treatment, doctors take a piece of the cancerous tumor from a patient, infect mice subjects with it, and test different drugs on the mice until they find a drug that shrinks the tumor. This treatment is then used on the human patient. Because the laboratory mice are bred to have the exact same genetic disposition as humans, their reactions to various medications can be judged more accurately. Also, these genetically engineered mice have no immune system, so their bodies do not reject the tumor. As lead researcher in the University of Colorado project Dr. Antonio Jimeno says, "We create a pseudo-patient." The benefits are clear, as he explains: "Patients from all walks of life understand—'You are testing drugs in mice, and when you find out which one works, you will give it to me. Then you won't have to test them all on me.'"[16]

Although this kind of research works for tumors of the head and neck, it cannot work for all tumors. Some tumors, such as those that form internally, are contaminated with other bacteria, and doctors could not get clear results in this "direct transfer" type of research. When it does work, however, it allows doctors to test a variety of drugs quickly and effectively.

"Animals cannot be adequately replaced. Substituting nonanimal methods for testing in . . . most medical research is a wishful fantasy."[14]

— Carl Cohen, professor of philosophy at the Residential College of the University of Michigan in Ann Arbor.

Why I Experiment on Animals

Researcher Steven Rose explains why he believes the use of animals in the laboratory is justifiable:

Let me be clear. I am an animal experimenter, a "vivisector" in the quaint parlance of those who oppose such work, although my research does not entail cutting up living creatures, as the term might imply. My experiments, primarily working with day-old chicks, are geared towards understanding the molecular processes that occur in their brains when they learn and remember new tasks.

For most of my researching life I have been prepared to justify this work as "basic science" aimed at answering a fundamental question about how the brain works. When opponents of such research asked what possible relevance does a chick brain have to a human brain, my answer would be to show them a chick and a human nerve cell side by side under the microscope. Not even the most skilled anatomist could tell them apart. . . .

As for the equivalence of pain, I don't find this a helpful concept. As a biologist, I accept that many species with complex nervous systems feel pain. Yet I find it difficult to imagine that a human faced with the choice of rescuing a chick or a baby from a burning house would hesitate. I strongly believe that precisely because we are humans, with our unique consciousness, we have responsibilities towards other forms of life. And overwhelmingly in my experience, animal researchers treat their subjects responsibly and with respect—it wouldn't be possible to study the behavior that I do if I regarded my chicks as little logic circuits, rather than living organisms.

Steven Rose, "Why I Experiment on Animals," *Guardian* (Manchester, UK), July 29, 2004. www.guardian.co.uk.

Advocates argue that such research is irreplaceable because mice and rats are very similar to humans both physiologically and biochemically. Mice and humans share major organ systems, enzymes, metabolic pathways, and over 98 percent of their genes. As writer Charles Alden says, "A few hundred rodents are exposed to a test chemical for the length of their normal lifespan, usually two years, to serve as surrogates for millions of people potentially exposed for decades."[17]

In fact, some researchers argue that mice have become a crucial corollary to humans for all medical research. Researcher Gary Wolf in *Wired* claims:

> With . . . experiments, you can go back to the lab and develop new tests and even therapies. Mice with their immune systems turned off can be reprogrammed with genetic code for human immune function to test treatments for diseases like AIDS. . . .
>
> The new lab mouse is no longer really a miniature human; it is a kind of genomic explorer that allows us to move back and forth between life and code. We don't know which new techniques of mouse science we will ultimately apply to ourselves. Maybe all of them.[18]

Wolf and others argue that it does not matter whether such research may result in a definitive cure. Simply by completing, repeating, and following such research, other patterns may become evident. Thus, unintended benefits of such research may arise that cannot be duplicated in any other way.

Man or Mouse?

Though these results clearly seem to benefit humans, detractors, including many physicians, argue that advocates of animal research grossly underplay the differences between humans and rodents such as mice and rats. In cancer research alone, many cancers, including those of the rectum and liver, are not reproducible in mice. In addition, many variables occur that are unpredictable and not reproducible in humans. In *Sacred Cows and Golden Geese: The Human Cost of Experiments on Animals*, C. Ray Greek, a medical doctor, and Jean Swingle Greek, a veterinarian, argue that scientists should "leave off curing mice and pick up the more pressing business of eradicating cancer in humans."[19]

The antiresearch camp states that, far from being uncommon, clinical trials in humans are where most "cures" come from. Drugs for breast cancer, such as tamoxifen, for example, were developed through clinical trials on women, not rodents. Dr. Richard C. Lewontin claims: "Most cures for cancer involve either removing the growing tumor or destroying it with powerful radiation or chemicals. . . . Medicine remains, despite all the talk of scientific medicine, essentially an empirical process in which one does what works."[20]

Because mice and rats are physiologically and biochemically similar to humans, many scientists consider them to be essential to biological and medical research. Pictured here are cages of laboratory mice being used in scientific research.

Thus, many physicians argue that animal experimentation is impractical and inefficient, and that results on animals rarely translate into superior treatments in humans. These medical experts argue that animal experimentation is not just bad for animals, but delays science and wastes research money by compiling useless data that are untranslatable to human patients. The Greeks explain:

> Animal experimentation has failed miserably, at tremendous expense, and has done real harm to human patients. True advances in medical knowledge have not come from animals. . . . Alternative methods—autopsies, *in vitro* research, clinical observation, epidemiology, mathematical modeling, and other human-based research modalities, could have resulted in the same achievements without injury to humans. Animal models are inaccurate, superfluous, and create risk to humans.[21]

The Greeks go on to argue that even though animal research consistently fails to replicate disease in humans and so fails in its conclusions, the business of animal experimentation continues. No new treatment or advance has been made using animals in cancer research.

Dr. Irwin Boss, the former director of Biostatics at the Roswell Park Institute for Cancer Research, agrees with the Greeks. He makes the point that animals are only used after scientists demonstrate that a particular scientific premise is worth proving. Once potential therapies are tested on animals, however, they are not necessarily applicable to humans. In fact, they may even harm humans. Boss says:

> From a scientific standpoint "animal model systems" in cancer research have been a total failure. . . . Not a single essential new drug for the treatment of human cancer was first picked up by an animal model system. All of the drugs in wide current clinical use were only put into animal model systems after finding clinical clues to their therapeutic possibility. The moral is that animal model

systems not only kill animals, they also kill humans. There is no good factual evidence to show the use of animals in cancer research has led to the prevention of or cure of a single human cancer.[22]

Use of Animals in Drug Development

Opponents of animal research use the development of drugs as a significant example of how animal research does not work. Medical protocol for drugs usually follows a common route: Test the safety and effectiveness of a drug on animals first, and if the drug works, proceed to human subjects. Thus, human subjects are always necessary for new drug development, discounting the theory that animal testing avoids testing on humans. In fact, drugs such as thalidomide and DES, tested successfully on animals, had disastrous results in humans, causing pregnant woman who used it to give birth to badly deformed babies. Thus, this protocol is fatally flawed, detractors argue.

What perhaps is most confusing in the animal testing debate is that advocates argue that even when animal testing fails to prove that a drug is safe for humans, such as with thalidomide, this does not prove the ineffectiveness of animal testing. Since no animal tests involved pregnant animals, the findings were delayed until the drug was used on humans.

Advocates argue that such research was still important, just flawed. The result was found to occur in animals as well, once pregnant animals were tested. The rush to get a drug to market was more at fault than animal testing. Thus, better oversight and peer review of research into drugs is needed, not an elimination of animals.

Most drugs must still go through human trials, even after animals are used. Rather than using failed, inefficient, and unreliable research on animals that ultimately must go through human trials anyway, the Greeks and others argue that it would be better to start with the human trials and come up with a better drug that is more targeted toward human use. But drug companies want a quick fix and to put something on the market as quickly as possible, even if

"Patients from all walks of life understand—'You are testing drugs in mice, and when you find out which one works, you will give it to me. Then you won't have to test them all on me.'"[16]

— University of Colorado cancer researcher Dr. Antonio Jimeno.

The Nobel Prize and Animal Research

For over a century the Nobel Prize has been awarded in recognition of pivotal medical and scientific discoveries. Of the more than 100 Nobel Prizes awarded for Physiology or Medicine, 75 have been based on some form of animal research; others relied on data previously obtained from biomedical research utilizing animal models. In 1905, for example, Robert Koch received the prize for his work with mice, rats, and guinea pigs that led to an understanding of the transmission and treatment of tuberculosis, a deadly, infectious disease that kills more than half of its victims if left untreated. Thomas Hunt Morgan received his Nobel Prize in 1933 after his experiments with fruit flies revealed the role of chromosomes in heredity, which helped pave the way for the modern science of genetics. In 1990 Joseph Murray and E. Donnall Thomas performed experiments on mice, rabbits, and dogs that led to the discovery of surgical techniques for organ transplantations that have saved the lives of thousands of people. These are but a few of the Nobel Prize–winning medical discoveries that have undeniably advanced human health. Many believe that animal experimentation remains the fundamental underpinning of this and other research.

it may ultimately be proved ineffective. What seems clear is that current medical research continues to rely on the animal model and that some of this research has led to the development of new procedures and drugs for humans.

Pain Research

One of the most controversial areas of medical research that involves animals centers on the genesis and treatment of pain. Currently, this type of research relies heavily on rodent models, although dogs, primates, and other species have been employed.

Many believe that these types of studies are crucial in that they help scientists better understand the mechanisms of pain and can inform better treatments for humans suffering from intractable chronic pain. At the same time, a large number of animal rights advocates view the intentional infliction of pain on a captive animal as cruel and unnecessary.

A 2009 study by psychology professors at McGill University and the University of British Columbia illustrates this wide divide. On one hand, the research findings may aid scientists in the understanding and treatment of human and animal pain alike: Using mice as subjects, the researchers inflicted pain by immersing the tail in hot water, cutting the paw, putting a binder clip on the tail, injecting chemicals into the stomach, and other procedures that induced pain that was comparable, according to the researchers, to that of a headache that could be relieved with analgesics such as Tylenol.

Researchers then videotaped the facial expressions of the mice and concluded that they express physical discomfort through facial expressions in the same way humans do. Based on these observations, the research teams developed the so-called Mouse Grimace Scale to measure pain responses by scoring five facial features—eye closing, nose and cheek bulging, and ear and whisker positioning—by which researchers could assess the severity of pain the mice were experiencing. According to Jeffrey Mogil, one of the lead researchers: "The Mouse Grimace Scale provides a measurement system that will both accelerate the development of new analgesics for humans, but also eliminate unnecessary suffering of laboratory mice in biomedical research."[23]

Others found this study highly unethical. Details about the study, which were first published in 2010 in the journal *Nature Methods*, elicited strong responses from those who took the view that the animals had been subjected to torture. Others took the view that the results were not particularly relevant to the body of research surrounding the genesis and treatment of pain; the infliction of pain on a sentient creature, therefore, was not justifiable. Clearly, the ethical implications of using animals in the search for effective treatment of pain will remain hotly debated for years to come.

"[Scientists should] leave off curing mice and pick up the more pressing business of eradicating cancer in humans."[19]

— Medical doctor C. Ray Greek and veterinarian Jean Swingle Greek.

Nutritional Studies

One of the most influential animal rights organizations, PETA, states on its website: "Medical historians have shown that improved nutrition, sanitation, and other behavioral and environmental factors—not anything learned from animal experiments—are responsible for the decline in deaths since 1900 from the most common infectious diseases and that medicine has had little to do with increased life expectancy."[24]

Many animal experimentation advocates agree with the premise of PETA's claim—that diet and advances in sanitation have largely improved public health—but assert that many of today's universally accepted beliefs about nutrition are, in fact, a direct result of animal research. Prior to 1900 vitamin and nutrition research was in its infancy. Animal models were used extensively during the next half century, when knowledge about the quality and quantity of essential nutrients needed by the body—vitamins, minerals, amino acids, fatty acids, and others—exploded.

A variety of debilitating diseases plagued large populations before the science of nutrition had been put forward to influence people's dietary habits. Scurvy, brought on by a deficiency in vitamin C, causes spongy gums, pus-filled skin wounds, and eventual death; rickets, due to a vitamin D deficiency, causes soft and malformed bones; and pellagra, caused by a deficiency of niacin B vitamin, produces a variety of symptoms, including skin disorders, dementia, and death. Animal models played an important role in the scientific understanding of these and other nutritionally induced diseases.

In a series of experiments in the early 1900s, Norwegian physicists Axel Holst and Theodor Frolich induced scurvy in guinea pigs. Although the symptoms of scurvy had been observed in humans, the guinea pigs with scurvy served as important models because they allowed, for the first time, the systematic study of the cause and cure of the disease. Holst and Frolich's findings—that scurvy could be prevented and treated through dietary substances—were published in 1907 and unleashed great debate in the scientific community because the idea of diseases being caused by nutritional deficiencies was not widely accepted at the time. Their groundbreaking experiment led to an understanding of the preventive value of different nutrients needed by the body and is today considered the most important single contribution to the understanding of scurvy.

In 1919 the British scientist Sir Edward Mellanby induced rickets in dogs, proving that it was a dietary deficiency that could be treated with cod liver oil, which is rich in vitamin D. As a result of these animal experiments, rickets is no longer a major public health problem. In the 1930s American biochemist Conrad Elvehjem and his colleagues studied dogs afflicted with black tongue, a disease analogous to pellagra in humans. By modifying the dog's diet, Elvehjem demonstrated that pellagra could be prevented and cured with niacin, one of the B vitamins. The results of these dog experiments had a direct impact on public health: In 1937 roughly 200,000 Americans had pellagra; 10,000 of these cases resulted in death. By 1940 there were fewer than 9,000 cases of this painful disease.

In another famous experiment from the early 1900s, American scientist Elmer McCollum conducted a series of nutritional experiments on rats and other small animals that led to the discovery of vitamin A—and the realization that millions of people were deficient in this important vitamin. The brutal consequences of vitamin A deficiency, which include congenital malformations, blindness, nerve injury, and eventually death, have largely been eliminated in the developed world, partially through the knowledge gained through these early animal experiments. To many, these and other examples from twentieth-century nutritional science highlight the essential role of animals in research. At the same time, animal rights activists assert that these medical gains could have come about had nonanimal research methods been employed. To date, animal research remains under way to assess how vitamins interact with other dietary components, the efficacy and safety of fortifying food products with particular nutrients, and other areas.

> "There is no good factual evidence to show the use of animals in cancer research has led to the prevention of or cure of a single human cancer."[22]
>
> — Dr. Irwin Boss, former director of Biostatics at the Roswell Park Institute for Cancer Research.

Xenotransplantation

For patients suffering heart defects, liver disease, kidney failure, or a variety of other diseases that cause organ failure, an organ trans-

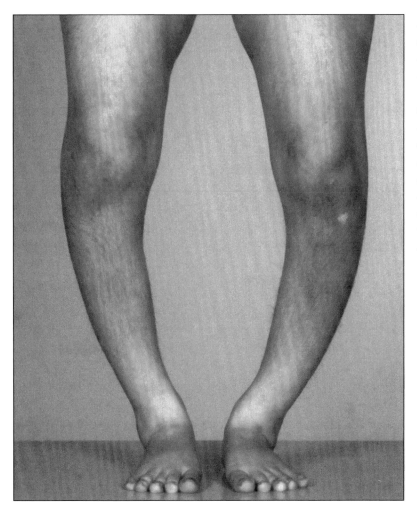

Experimentation on dogs in the early 1900s led to discovery of the cause of rickets, which is caused by a deficiency of vitamin D or calcium and lack of exposure to sunlight. Rickets occurs chiefly in children and leads to a weakening of the bones, which causes various deformities, including bowed legs.

plant is often their only hope of survival. Doctors have made great strides in the technology of transplantation, and today human-to-human organ transplants are common and have saved the lives of many otherwise untreatable patients. At the same time, the number of available healthy organs from human donors is limited, and many people die while on a waiting list to receive a lifesaving organ. According to the US Food and Drug Administration, 10 patients in the United States die each day while waiting for an organ.

A number of doctors and scientists believe that xenotransplantation—the transfer of living cells, tissue, or organs from one species to another—holds the key to addressing the chronic shortage of available organs. Primates and pigs appear to be the most suitable organ donors. Primates were first considered as the favored organ source since they are the closest relatives to humans, with strikingly similar anatomies.

Using animal tissue and organs from these animals is controversial. To many the idea of sacrificing animals for their organs to treat human conditions raises a number of ethical issues. Many find the use of primates in this endeavor especially troublesome, as they view these highly sentient animals as having inherent rights similar to those of humans. Although the use of pigs obviates some of these ethical concerns, many animal rights groups oppose the use of any animal species. According to PETA:

> Since the 20th century, dozens of pigs, chimpanzees, monkeys, and baboons have been made the unwilling "donors" of kidneys, hearts, livers, and bone marrow for transplantation into humans. Additionally, thousands of goats, rats, chickens, cats, and dogs have died in cross-species experimentation. . . . These animals are subjected to sensory deprivation in sterile laboratory environments and are denied social interaction with members of their own species. One investigation of a British research facility revealed the torturous lives and deaths of animals subjected to xenotransplantation. Monkeys and baboons died "in fits of vomiting and diarrhea . . ." and "other animals retreated within themselves, lying still in their cages until put out of their misery. . . ." Every one of these experiments has failed—most transplant recipients die within a few hours, days, or weeks of surgery.[25]

Others assert that human needs trump the rights of animals and that xenotransplantation can potentially save millions of human lives. Research addressing the feasibility of xenotransplantation has been under way for years. In 1963 and 1964 Dr. Thomas Starzl, one of the pioneers of human-to-human organ transplantation, performed 6 baboon-to-human kidney transplantations; Starzl also performed the first baboon-to-human liver transplants in 1992 and 1993. Although the patients all died shortly after their surgeries due to infections associated with immunosuppressive drugs, many believe that these experiments laid the groundwork for ongoing research that will one day expand the availability of organs to dying patients.

Barriers to Overcome

Many barriers must be overcome before xenotransplantation can preserve human life. For example, whether or not animal organs can fully replace the physiological functions of human organs has yet to be determined. The transmission of animal diseases into the human population is another concern. Baboons, pigs, and other animals harbor many transmittable viruses, bacteria, and fungi that, while harmless in the animal host, are deadly to humans. Retroviruses are especially troublesome: The HIV retrovirus, for example, is fatal to humans but relatively harmless in monkeys. Scientists speculate that HIV may have originated in nonhuman primates and that xenotransplantation could unleash new epidemics of similarly infectious diseases into the human population.

Perhaps the most formidable obstacles to overcome are posed by the human recipient's immune system, which is primed to reject any foreign matter in the body—bacteria, viruses, animal organs, or even artificial body parts. Even when patients receive organs from human donors, it is a medical challenge to keep their immune systems from rejecting them. The immunological response to animal organs is even more extreme. Thus far, no xenotransplantation trials have been successful due to the response of the recipient's immune system.

Genetic engineering may hold the key to overcoming organ rejection. Researchers are studying ways to insert human genes into animal organs so that the body will identify them as "human" and thereby minimize the risk of rejection. Gene therapies, however, are still in the early stages of development. Many scientists believe that until the science progresses, transplantations involving cells and tissues such as bone marrow, as opposed to whole organs, are more realistic procedures in the short term. Whatever the case, it is likely that the scarcity of human organs will continue to prompt intense research that employs animal models as scientists attempt to overcome the roadblocks to successful xenotransplantation. Whether medical gains in this area can be achieved without the use of animal models remains an interesting but unanswered question.

Facts

- According to the US Department of Agriculture, the total number of animals used for experimentation in the United States was over 1 million in 2007. This figure does not include rats and mice.

- Class A dealers are licensed by the US Department of Agriculture to sell animals that are purposely bred for research and educational purposes.

- Although there are no exact figures available, an estimated 20 million rats and mice are used in experiments each year in the United States, making these rodents the most commonly used vertebrate species.

- In February 2011 the journal *Nature* published results of a poll of 980 biomedical scientists around the globe, reporting that 91.7 percent said they strongly agreed that animal research was essential for the advancement of biomedical science.

- Federal law requires that all drugs and some chemicals, particularly those used in the production of food, be tested on animals before humans.

- The Johns Hopkins University Center for Alternatives to Animal Testing was established in 1981 to find alternatives to animal testing that would not impede medical progress.

Should Primates Be Used in Research?

Humans have long known of the evolutionary history that they share with nonhuman primates. Celebrated chimpanzee researcher Jane Goodall, mountain gorilla expert Dian Fossey, and other researchers have driven home the similarities in behavior, family closeness, and intelligence that humans share with these creatures. Just one example is Koko, the mountain gorilla who has captivated people with her language skills. Koko knows more words than any other nonhuman. Using American Sign Language, Koko has a vocabulary of over 1,000 signs and understands approximately 2,000 words of spoken English. She can put together sentences of 6 to 8 words. Even more interestingly, Koko often strings together new words and phrases based on her vocabulary. With an IQ of between 70 and 95, Koko was close to the average human IQ of 100. And it is not just her intelligence that is comparable to humans'. On video, Koko endearingly initiates conversations with her handlers, holds, hugs, and blows kisses to a doll, and shows other humanlike behaviors.

Mountain Gorillas

Fossey, the researcher made famous by the movie *Gorillas in the Mist*, lived side by side with gorillas in the wild for many years, spending whole days in their habitat and imitating their behavior. This intimate contact revealed the close family relationships and bonds gorillas have with each other.

Mountain gorillas rest in the jungles of Rwanda in Africa. Mountain gorillas are intelligent creatures and, like humans, have close family bonds and relationships. Awareness of the similarities between the great apes and humans has influenced the debate over their use in scientific research.

Fossey and Goodall revealed much of their research to the public to protect the habitats of the great apes. They wanted humans to become aware of the similarity of behaviors, feelings, needs, and desires between humans and apes. They believed that this would engender empathy and move people to participate in fund-raising, ecotours, and protests to save the gorillas' native habitats.

These emotional appeals to understand our closest relatives have indirectly influenced the debate over whether nonhuman primates should be used in medical research. In the case of the monkeys of Silver Springs, Maryland, scientists and animal rights activists both found support for their side of the debate over primate research. In 1989 researcher Edward Taub, a psychologist, was conducting a long-term study on neuroplasticity, or the ability of the brain to remap itself and allow undamaged areas to take over for damaged ones. The research involved 17 macaque monkeys; researchers had cut the ganglia that connected the monkeys' arms and legs with their brains. Once the connection from the brain to the limbs had been severed, Taub used restraints, electric shock, and withholding of food to force them to use the limbs they could not feel.

The Silver Springs Monkeys

In the summer of 1981, Alex Pacheco of PETA, an animal rights group, infiltrated the Silver Springs lab by working undercover. He took pictures of conditions in the lab and alerted the police that the monkeys were living in unsanitary conditions, with feces stacked so high it entered the cages, and food and water bowls fouled with feces. The monkeys were seized by police, and survivors (many have died in the meantime) remain in legal limbo to this day. The court case has never been resolved, and the monkeys are neither the property of Taub nor anyone else. Taub, originally charged with 113 counts of animal cruelty, was later cleared of all charges.

When the dead monkeys were later dissected, they showed remarkable brain remapping, proving Taub's theory of neuroplasticity. Originally maligned and unable to find a job, Taub was lauded and received a grant at the University of Alabama. There, he developed treatments for brain-damaged individuals. This therapy has helped stroke survivors regain the use of their limbs.

The Silver Springs monkey case has become a rallying cry both for animal research detractors and supporters. Pacheco argues that the jury verdict was based on an incomplete picture of how the monkeys were treated. He states:

> There were many things that the jury was never allowed to consider in making its decision, things it was never allowed to hear, know about or see. For example, the jury was not permitted to hear about the discovery of two 55-gallon barrels filled with the corpses of monkeys and weighted down with used auto parts and wood. The jury could not ask, "What became of them? How did they die?" The jury was never allowed to hear that Taub was denied a grant application because between 80 and 90 per cent of his animal subjects died before the end of his experiments. It could not see the 1979 US Department of Agriculture inspection report that read: "Floors were dirty with blood stains all over them." It was never allowed to know that Taub operated illegally, in violation of federal law, for seven years, while receiving hundreds of thousands of federal tax dollars. The jury did not know that [a monkey named] Caligula suffered

from gangrene and mutilated his own chest cavity, that blood splattered the wall and ceiling of the converted refrigerator chamber, that the [National Institutes of Health] had investigated Taub and found him in violation of its own guidelines, that [a monkey named] Charlie had died of an unexplained "heart attack." It was never allowed to see or hear of the surgically severed monkey hand or the skull that Taub used as paperweights in his office. And, perhaps most unfortunately, the jury was never allowed to see the living evidence, the monkeys themselves.[26]

For Pacheco and other animal rights activists, the cruel conditions to which the monkeys were subjected can never be justified, no matter what benefits accrue for humanity.

Pioneering Treatments

On the other hand, others argue vociferously that the case proves the inanity of animal rights groups, their antiscience stance, and how their feelings for animals get in the way of understanding crucial research. Researchers argue that after their seizure from the lab, the monkeys were inadequately cared for, no one knew how to treat their injuries, and many died. Those monkeys that died without being autopsied made their sacrifices for nothing, the medical community argued. Those that were autopsied led to Taub's pioneering treatment of stroke victims known as constraint-induced movement therapy, which has been lauded by the American Stroke Association. In an article for *National Review*, Wesley J. Smith articulates the argument:

> The animal research that so distressed animal liberationists helped Taub achieve a medical breakthrough in the treatment of stroke victims—called Constraint-Induced Movement (CI Therapy)—by which the brain is induced to "rewire itself" following stroke or other serious brain trauma. CI Therapy is so successful that there is now a long waiting list of stroke patients with upper limb impairments at Taub's Alabama clinic. The technique is also in further human trials for other conditions, including as

an approach to treating children with cerebral palsy and traumatic brain injury.

It is frightening to think that if Pacheco had successfully ruined Taub, CI Therapy might have been lost to humanity. "We are contrasting the treatment of thirteen monkeys with the improved motor ability and quality of life of thousands of human beings," Taub told me a few years ago. "If I had been unable to continue with my research, it would have left the burden of thousands of stroke victims unalleviated."[27]

This argument, that monkeys are invaluable research subjects, is why these animals continue to be used. As the California Biomedical Research Association argues: "Because nonhuman primates reflect the anatomical, physiological, and behavioral makeup of humans, they provide an indispensable, and currently *irreplaceable*, bridge between basic laboratory studies and clinical use. Much of what we know about the brain, heart disease, Alzheimer's, AIDS, viruses, hepatitis, and cancer has come from nonhuman primates."[28]

Genetic Similarities

It is hard to deny the genetic similarities between humans and nonhuman primates. Human DNA and that of chimpanzees, for example, is more than 98.3 percent identical. Human and chimpanzee brains, too, are strikingly similar. For example, 80 percent of the human brain and 75 percent of the chimpanzee brain is cerebral cortex. The most forward section of the cortex is the frontal lobe, which is often associated with the most complex mental activities—artistic expression, language, creative thinking, and some aspects of emotional behavior, for example.

Commenting on the complexity of the nonhuman primate brain and its similarity to that of the human brain, Frederick King, former director of the Yerkes National Primate Research Center, notes that "the large, convoluted cerebral cortex, with great areas devoted to associational activities, is almost certainly responsible

"Much of what we know about the brain, heart disease, Alzheimer's, AIDS, viruses, hepatitis, and cancer has come from nonhuman primates."[28]

— California Biomedical Research Association.

Deprivation Studies

British primatologist Jane Goodall and others have observed that baby chimpanzees and other primates are dependent on their mothers and display behaviors that are very similar to human babies. American psychologist Harry Harlow conducted a series of maternal separation experiments on monkeys in the 1950s and 1960s. His experiments were controversial because some of them involved rearing monkeys in partial or total isolation, from which they emerged psychologically disturbed. His research also confirmed striking similarities between humans and monkeys, as he writes:

> The macaque infant differs from the human infant in that the monkey is more mature at birth and grows more rapidly; but the basic responses relating to affection, including nursing, contact, clinging, and even visual and auditory exploration, exhibit no fundamental differences in the two species. Even the development of perception, fear, frustration, and learning capability follows very similar sequences in rhesus monkeys and human children.

> While Harlow's studies are generally considered unethical by today's standards, many believe that he provided valuable insight into the nature of love, emotional attachment, and loss. Paradoxically, Harlow heightened awareness of how animals are treated in captivity and may have been a driving force behind today's animal protection movement.

Harry Harlow, "The Nature of Love," *American Psychologist*, vol. 13, 1958. psychclassics.yorku.ca.

for the primate's ability to learn highly complex cognitive tasks beyond the capacities of species other than humans."[29]

Indeed, primates' psychological and physiological similarity to humans is clear; it is unclear, however, just how many primates are used in research each year because universities, government-funded facilities, and private facilities do not always report accurately how many animals are used. According to a 2010 USDA report, close to 70,000 primates were used in research in 2009 in the United States. Many of these primates are rhesus macaque monkeys, but squirrel monkeys, chimps, marmosets, and others are used as well. Almost all of these primates are bred in captivity. They are bred at 8 National Primate Research Centers and other institutions in the United States.

Recent Primate Research

In the first decade of the 2000s alone, scientists attributed many medical advances to research conducted on primates. Some of the most important include research into HIV/AIDS, Parkinson's disease, and hepatitis B and C.

Monkeys can carry a strain of virus very similar to the HIV virus that follows the same progression as the human counterpart. Because of this, monkeys are important in testing developmental vaccines for the disease. Other research with primates includes exploring why intravenous drug users are more likely to become infected with HIV and also how maternal transmission of the virus to infants is conducted.

Due largely to the use of monkeys in research, scientists have virtually eradicated hepatitis B and C infections from blood transfusions. Chimpanzees have been the primary subjects in hepatitis research because they are the only animals besides humans that are susceptible to this viral infection, which can cause chronic liver disease, cirrhosis, and liver cancer. Research commenced in the early 1970s, when a group of American scientists founded Vilab, a biomedical research institute in Liberia, primarily to study hepatitis B and C, two particularly virulent forms of hepatitis. Establishing the center in Liberia meant that researchers would not have to import wild-caught chimpanzees to the United States. Largely through this research, a hepatitis B vaccine was made available in 1982 and is protecting millions of people from cirrhosis and liver cancer throughout the world today. Vilab scientists also created blood sterilization techniques that have virtually eradicated the spread of hepatitis B and C, as well as the AIDS virus, through blood transfusions. Efforts to develop a hepatitis C vaccine have been unsuccessful thus far, although research is ongoing.

In Parkinson's research, scientists inject a gene known to protect brain cells directly into the brains of monkeys. This experimental treatment slowed the progression of the disease, and scientists are hoping to move to human trials soon. Professor Roger Morris of King's College London commented on this research: "We should not forget that the few experiments we carry out on primates have the potential to alleviate a vast amount of human suffering. In the case of Parkinson's a few thousand animals will help develop models that could prevent hundreds of thousands of people suffering lingering deaths, their brains etched from within, and whose families face terrible emotional suffering."[30]

The argument is a potent one: Without the use of primates, those on the side of research contend, there are only two alternatives. Researchers can cease efforts to develop crucial medicines and treatments that only infect primates, or they can conduct tests on human subjects, which would be appallingly inhumane.

Condemning Experimentation on Primates

Detractors take issue with both of these dire predictions. Not only are primates *not* indispensable but relying on the results of research gained from experimenting on them to help humans is rarely productive. Detractors argue that in almost every case, experimentation on primates results in false conclusions and treatments that cannot translate to human needs.

As Jarrod Bailey, science director for Europeans for Medical Progress, contends in an article published in *Biogenic Amines*, "Many papers reveal findings that, while constituting interesting contributions to the scientific knowledge-base, are not relevant or applicable to human medicine."[31] Indeed, just about every area in which proponents claim progress due to research in monkeys and apes has its detractors.

In the examples above, researchers claim that not a single instance of primate-related research has led to positive results. Parkinson's disease in humans, for example, largely attacks the elderly. Its symptoms are characterized by lesions on the brain, and it takes years for the disease to develop and debilitate humans. The course of the disease in monkey research, however, by necessity must be conducted in a much smaller time frame due to monkeys' shorter life span, and many of the ways the disease develops in humans are not found in monkeys. Most major breakthroughs in Parkinson's have been made in clinical studies, genetic research, and human tissue studies and autopsies.

As another example, the AIDS virus in monkeys is distinctly different than the virus in humans. As Bailey writes: "Everything we know about HIV and AIDS has been discovered without relying on animal models. Effective HIV protease-inhibitor and nucleoside-analogue drugs were conceived and developed using *in vitro* and *silico* methods."[32] C. Ray Greek and Jean Swingle Greek issue an even harsher condemnation: "How many ways is animal experimentation for HIV unsound? It is needless, inaccurate, unreliable, repetitious, and improperly motivated. It makes special-interest groups rich while people become sick and die."[33]

The argument that primate research is simply not useful is dependent on the idea that infectious diseases, even when they can be duplicated in a primate, usually do not follow the same course, may or may not be as virulent, and may or may not behave in the same ways in humans. Bailey sums it up: "On a superficial level we don't look alike or behave similarly; on a deeper level our biochemical differences mean that we suffer from different diseases, respond in different ways to infectious agents, have different metabolisms, and find different substances toxic. In short, an awful lot of subtle biochemical differences combine to make us very different indeed."[34]

> "Many papers reveal findings that, while constituting interesting contributions to the scientific knowledge-base, are not relevant or applicable to human medicine."[31]
>
> — Jarrod Bailey, science director for Europeans for Medical Progress.

Primate Research Continues

Despite such views, research on primates remains under way at the nation's eight primate centers, universities, and other centers of animal study. For example, the California National Primate Research Center just released promising results from spinal cord injury research being conducted on rhesus monkeys showing the generation of new nerve fiber growth after a spinal cord injury. According to the center: "This phenomenon has not been seen in the rat—the traditional research model for spinal injuries. The work highlights

Young monkeys huddle together at the Tufts University veterinary school. Despite the similarities between human and nonhuman primates, some experts say, infectious diseases affect both quite differently—which greatly reduces the value of primate research.

an important role for primate models in translating basic scientific research into practical, therapeutic treatments for people. The spinal cords of humans and other primates are different from rodents, both in overall anatomy and in specific functions."[35]

Results from experiments like these underscore the difficulties in assessing the role of primates in research. Like Taub's monkey studies, this type of research offers hope to humans with incapacitating spinal injuries. At the same time, many question the ethics of subjecting primates—highly sentient animals with complex emotional and cognitive lives—to painful spinal cord procedures.

Protections for Nonhuman Primates

Primates are protected by many of the same laws and regulations that guide all animal experiments. The American Society of Primatologists is a nonprofit organization that promotes the study and conservation of primates and their use as research models. The group's website describes some of the regulations that pertain specifically to primate research:

> Because nonhuman primates are highly regulated in the United States, any experiment that a scientist proposes to conduct with monkeys must be approved by the Institutional Animal Care and Use Committee (IACUC) at the institution where the scientist works. The scientist must describe in detail the specific procedures that he or she plans to use on the animals, such as any behavioral testing, surgical procedures, or experimental substances like drugs or vaccines that the animals might receive. There must be an explanation of whether any of the procedures are likely to cause the animals pain or distress, and if so, details must be presented describing all steps the scientist will take to minimize or eliminate pain or distress. The scientist must also provide a justification for why the proposed research must be conducted with monkeys rather than some other animal; whether there are any alternative ways that the scientist can find the answer to his or her question.[36]

Other individuals and organizations advocate a complete ban on primate testing. One prominent movement is the Great Ape Project (GAP), launched in 1993 following the publication of a book edited by Peter Singer and Paola Cavalieri that highlighted the similarities between humans and primates, such as social organization and the ability to form strong familial bonds. The primatologists, anthropologists, ethicists, and others who constitute GAP advocate a United Nations declaration of the rights of great apes based

Social Housing

Research facilities receiving federal funds are required to follow the *Guide for the Care and Use of Laboratory Animals*. The following excerpt from this handbook defines the parameters that govern a primate's housing and social environment:

> Consideration should be given to an animal's social needs. The social environment usually involves physical contact and communication among members of the same species (conspecifics), although it can include noncontact communication among individuals through visual, auditory, and olfactory signals. When it is appropriate and compatible with the protocol, social animals should be housed in physical contact with conspecifics. For example, grouping of social primates or canids [family of mammals that includes foxes, wolves, and dogs] is often beneficial to them if groups comprise compatible individuals. Appropriate social interactions among conspecifics are essential for normal development in many species. A social companion might buffer the effects of a stressful situation, reduce behavioral abnormality, increase opportunities for exercise, and expand species-typical behavior and cognitive stimulation. Such factors as population density, ability to disperse, initial familiarity among animals, and social rank should be evaluated when animals are being grouped.

Institute of Laboratory Animal Research Commission on Life Sciences, National Research Council, *Guide for the Care and Use of Laboratory Animals*. Washington, DC: National Academy Press, 1996. http://olaw.nih.gov.

on the group's charge that the sentience which great apes display affords them greater legal protections—essentially an extension of rights usually only granted to humans.

According to the group's website, "GAP is an international movement that aims to defend the rights of the nonhuman great primates—chimpanzees, gorillas, orangutans and bonobos, our closest relatives in the animal kingdom. The main rights are: the right to life, the protection of individual liberty and the prohibition of torture."[37]

The GAP project is affiliated with four chimpanzee sanctuaries; most of the chimps protected in these wildlife preserves were rescued from circuses and zoos, and a smaller minority from animal research facilities. Many of the chimps that make their way to these sanctuaries must first be treated for physical disorders related to living in captivity. Many animal handlers have observed that being caged or isolated is uniquely stressful to chimpanzees and other primates, which exhibit a variety of reactions to social deprivation, including rocking back and forth and appearing tired and listless. In more extreme instances, these animals have been documented as plucking out their fur, banging their heads against the cage, and even gnawing at their limbs.

Dr. Pedro A. Ynterian, director of GAP International since 2006, concludes that "a chimpanzee is not a pet and cannot be used as an object for fun or scientific experiment. He or she thinks, develops affection, hates, suffers, learns and even transmits knowledge. To sum it up, they are just like us. The only difference is that they don't speak, but they communicate through gestures, sounds and facial expressions."[38] To Ynterian and others, this similarity to humans renders the use of the great apes in scientific laboratories or the entertainment industry akin to slavery.

Clearly, the use of primates in animal research remains in debate. The question of whether these animals' biological and emotional similarities yield valid results that apply to the treatment of human disease is answered by an absolute yes from some and an equally emphatic no from others. Today research with primates continues to be a large part of animal experimentation.

"How many ways is animal experimentation for HIV unsound? It is needless, inaccurate, unreliable, repetitious, and improperly motivated."[33]

— Medical doctor C. Ray Greek and veterinarian Jean Swingle Greek.

Facts

- In June 2008 Spain's parliament passed a resolution that could give certain primates, including chimpanzees, bonobos, gorillas, and orangutans, legal protection from animal experimentation.

- Each year over 30,000 wild primates are captured and sold on the international market. The United States imports close to one-third of these primates, primarily for use in research.

- The Convention on International Trade in Endangered Species sets forth regulations governing the international trade in primates. More than 120 countries have agreed to abide by these guidelines.

- In 2006 a primate researcher at the University of California–Los Angeles shut down a series of experiments that involved macaque monkeys after receiving death threats from animal rights activists. Similar incidents led to the passage of the Animal Enterprise Terrorism Act to protect animal research organizations.

- Primates are often used to study issues related to reproductive health because their menstrual cycles and neonatal development are similar to humans'.

- Many of the individual primates that are used in biomedical or other experiments are used in multiple studies over the course of many years.

- According to the American Society of Primatologists, the majority of primates studied in the United States are bred in one of several US facilities for use in research, testing, or education.

Animals in Education

Jennifer Graham was a 15-year-old student in Victorville, California, when she refused to dissect a frog in her high school biology class. Graham argued that she wanted no part in harming and killing an animal for the sake of her education. She told the teacher that she would do extra coursework to make up for her refusal. The teacher responded by dropping her A to a D, later upgrading it to a C. The teacher argued that dissection was an essential part of the course that could not be replaced by other work. Although the case went to court in 1991, it remained unresolved. The case did highlight, however, the issues that remain regarding dissection—or any type of use of animals—in the pursuit of education, whether in the high school or the college classroom.

Dissection in the Classroom

Animal dissection—exploratory surgery on an animal for educational purposes—has been part of biology education in the United States since the early 1900s. For millions of high school students across the country, dissecting a fetal pig, mouse, or other dead animal has become a rite of passage. To date, it remains unclear how many animals are used in American classrooms each year. According to the organization TeachKind, a group that promotes the use of human educational materials and alternatives to dissection, an estimated 20 million animals are used for educational purposes each year; about half of these are slated for dissection. Other groups put the figure much higher. A reasonable estimate is that as many 6 million vertebrates—commonly frogs, mice, rats, rabbits, cats, and fetal pigs—are dissected each year in US high schools alone. The number of invertebrates—earthworms, clams,

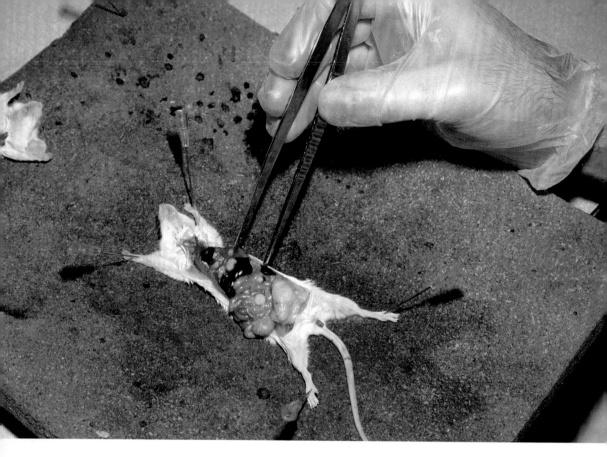

squid, insects, and others—is likely comparable. An additional, unknown number—perhaps millions—are used in colleges and middle and elementary schools.

Collection

Animals used in dissection are usually either captured in the wild, purchased at animal shelters, bred in facilities specifically to be dissected, or procured from slaughterhouses. Most of these animals are then processed through a biological supply company before they are sent to the classroom. The procurement industry is very profitable. In the United States alone, it is a multimillion-dollar industry. Animals that are taken from their natural habitat to be used in dissection include frogs, the most commonly dissected animal below the postsecondary level. Each year millions of frogs are caught in the wild, which adversely affects their natural habitat—by increasing insect populations, for example—and depletes certain species, which may become endangered.

In addition, few rules apply regarding the capture and transfer of such animals, and the methods are cruel—and usually hidden from public view. Animal advocate and author Jonathan Balcombe describes the treatment of frogs at one biological supply house:

> At the time of capture, frogs were kept in large sacks or cages. As many as 100 frogs were kept in each sack for up to a week or more, the only care being intermittent spraying with water. Eventually, the frogs were put into large tubs of water where they were kept for periods ranging from days to months depending on the season and the demand for shipments. During this period, the frogs were provided no food. Frogs shipped during the summer likely had gone without food for a week or more between capture and arrival at a school; in the early spring, frogs may not have eaten for more than six months. Live frogs were usually shipped 50 to a box lined with sphagnum moss. In the summer months, most frogs were "hot," meaning that they were overheated and hyperactive often to the point of convulsion.[39]

Higher-level animals do not fare much better. In his book *Empty Cages*, Tom Regan describes a video that was secretly filmed by an animal rights activist who infiltrated an animal research supply house:

> The video shows cats arriving at a biological supply company crammed so tightly into crates that they cannot stand up. Some are visibly sick; others look as if they are dying. Then it's on to the gas chamber. Many of the cats are still moving when workers pump formaldehyde into their veins. They clench their paws as the chemicals surge through their bodies. They are then stored and packaged and eventually shipped to schools all around the country.[40]

Whether these incidents are unusual is difficult to determine, since biological supply companies are rarely investigated, and their practices remain somewhat secretive. However, discussion remains over whether the benefits of dissection outweigh the huge cost to animal populations. Even supporters of dissection would no doubt find such cruelty reprehensible. Yet few think of the methods that must be used to supply the large number of animals required by schools and colleges.

Invaluable Tool?

Those who support the practice of dissection in the classroom assert that it provides critical hands-on experience to students learning about anatomy and physiological functions. In short, it allows students to view the internal structures of animals and to study how the organs and other tissues are interrelated. For those entering the medical and veterinary professions, dissection provides contact with real tissues— bones, veins, and internal organs. Such contact is essential to understanding surgical and other medical procedures.

Even for those students who are not entering a medical career, the experience of live dissection can provide a sense of wonder and respect for life by exposing students to the intricacies of live processes. High school biology teacher Susan Offner describes her first—and highly memorable—dissection experience:

I can still remember my first dissection of a mammal. It was a mouse. What ensued was a tremendous explosion of consciousness and understanding. All the things I had been learning were suddenly real. It was a profound experience. But it was something more. By confirming all the things I had been taught, it helped me understand that the world was a rational place, and that knowledge and understanding can come from serious study of real specimens and real data. I see this same kind of learning in my own students.[41]

Others recognize that dissection must be part of their training, especially if they are training for a role in medicine. Author Linda Birke describes a class of medical students after they had dissected dogs for an anatomy lesson:

"I can still remember my first dissection of a mammal. It was a mouse. What ensued was a tremendous explosion of consciousness and understanding."[41]

— Biology teacher Susan Offner.

Where before the lab there had been much trepidation and doubt, afterwards there was fascination and enthusiasm—just as several younger students reported after doing their first dissections. But for these older students, there is usually no opportunity to opt out—they have already chosen their professional path and have to get through the experiences. For them, fascination is a near universal response.[42]

Birke and her colleagues report the comments made by medical students regarding dissection:

"This was really neat—digging in and seeing the stuff actually working and pumping—it was great." This was "better than books." As another student pointed out, seeing things illustrated by models was less useful because they were never living, and because the students themselves did not change anything: "To me a model wouldn't be much different than a book. Nothing you do is going to change anything with the model."[43]

Psychological Impact on Students

While reactions to dissections vary, the practice provokes considerable discomfort in a number of students. Many express reservations about working with animals that were sacrificed for educational gains. Some researchers have reported that to carry out dissections, students must repress their natural feelings of disgust or repulsion so that they can cut up live, or once live, animals. Many worry about the negative psychological effect this may have on students. The worry is that dissection may desensitize students to animal suffering, which may foster callousness or even cruelty to animals.

As noted animal advocate and chimp researcher Jane Goodall puts it:

This type of education subjects the young people of our society to a kind of brainwashing that starts in school. . . . By and large, students are given the implicit message that it is ethically acceptable to perpetuate, in the name of science, a variety of unpleasant procedures against animals. They are encouraged to suppress any empathy they may feel for their subjects, and persuaded that animal pain and feelings are of a different nature from our own, and that there is little value in animal life.[44]

Or, as the zoologist Miriam Rothschild notes, "Just as we have to depersonalize human opponents in wartime in order to kill them with indifference, so we have to create a void between ourselves and the animals on which we inflict pain and misery."[45]

Dissection may be especially troubling when the animals being dissected are traditionally considered companion animals, such as dogs or cats. Living, anesthetized dogs, for example, are routinely used by medical schools to teach orthopedic and surgical techniques and veterinary medicine. After the work is completed, the animals are killed, often because an autopsy is necessary to see the results, or because the animal is too damaged to be kept alive. As one student reported: "I had a really negative expectation; I almost didn't go. I've had a lot of dogs. Dogs were always really important to me. I hunted with them. I trained them. They were a lot of times my best friends, like my dog Sam. I ran with him every morning. So I felt it would be really hard for me to see a dog in that type of setup. . . . I agonized over it quite a bit the week before."[46]

As another biologist put it: "In my view, undergraduate animal experiments often inflict unnecessary suffering on animals; they also have a hardening and desensitizing effect on the students required to perform them at a time when the development of a sympathetic attitude toward the natural world may be just as important as the teaching of actual scientific knowledge."[47]

Still, few would argue that for those students who choose to enter either veterinary or medical training, such desensitization is a necessary requirement for being able to act objectively to treat an animal or to become a surgeon. Dissecting living animals is invaluable for learning such things as using a scalpel on living tissue, keeping vital signs steady, and other procedures that must be mastered by a physician.

"By and large, students are given the implicit message that it is ethically acceptable to perpetuate, in the name of science, a variety of unpleasant procedures against animals."[44]

— British primatologist Jane Goodall.

Live Animal Labs

While the majority of medical schools in the United States have abolished dog labs to train students of medicine, a handful of schools continue to defend the practice of using live animals. At the Johns Hopkins University School of Medicine, for example, medical students spend two days working on live pigs—learning to stop bleeding, remove organs, and perform a variety of surgical techniques. Dr. Julie Freischlag, director of surgery at the school, asserts that pigs give the students the feel of live tissue and help determine whether they have the skill and dexterity to perform delicate surgical operations: "Simulators have no feedback as to texture and touch. That's why it's so important to use animals, to feel all the right tensions and strengths."

At the University of Washington, students who are learning to care for premature infants practice inserting breathing tubes into anesthetized ferrets. According to Dr. Dennis Mayock, medical director of the university's Neonatal Intensive Care Unit, training students with plastic models or other simulators is insufficient: "The airway tissue in extremely small infants cannot be duplicated or simulated well enough with any of these current models." The medically fragile infants that his department treats, Mayock asserts, would suffer irreversible brain damage or even die if doctors could not quickly and accurately intubate them.

Quoted in Jonathan Bor, "M.D. Group Protests Hopkins' Use of Pigs," *Baltimore Sun*, March 27, 2008. www.baltimoresun.com.

Quoted in Carol M. Ostrum, "Group Faults UW Use of Ferrets in Medical Training," *Seattle Times*, February 9, 2011. seattletimes.nwsource.com.

Giving Students and Teachers Choices

In the undergraduate schools, however, student concern about dissection has prompted many states—including Florida, California, Pennsylvania, New York, Rhode Island, Illinois, Virginia, Oregon, New Jersey, and Vermont—to enact "student choice laws." These laws allow students to opt out of dissection without being penalized. Other states and independent school districts have opted for similar policies that allow some form of student choice. These recent legislative moves ensure a student's right to use an array of viable alternatives to dissection.

In addition, in 2008 the National Science Teachers Association, an organization that supports the study of science in schools, formally stated in its mission statement that it supports a teacher's decision to use animal dissection as a teaching tool but also encourages nonanimal teaching methods.

Alternatives

Alternatives to dissection are abundant and diverse. In many classrooms today students learn life sciences using a multitude of videotapes, anatomical charts, peel-away transparencies, and three-dimensional plastic animal models with removable organs, for example. Three-dimensional CD-ROM presentations and sophisticated computer simulation programs provide many opportunities for interactive virtual dissections. The digital frog interactive CD-ROM, for example, allows students to "dissect" a computer-generated frog with a digital scalpel. Other accurate and easily accessible educational tools available include McGraw-Hill's Anatomy and Physiology REVEALED, a computer simulation that allows students to view human brain, heart, and circulatory functions, and Stanford University's Virtual Creatures, which allow students to interact with high-resolution computerized organisms.

One of the benefits of nonanimal dissection kits and other high-tech tools is that they allow for repeated use according to a student's needs, and they can be used by multiple students. In contrast, a dissected animal is used only once before it is discarded, leading many to condemn the practice as wasteful and impractical.

While most educators find dissection kits, computer programs, and other tools useful adjuncts to dissection, many believe that they cannot fully substitute for the actual experience of dissection. As Birke writes: "Cyber-dissection or other alternatives do not (yet) teach students the kind of corporeal [hands-on] skills that have long typified biology and how it is taught. You simply cannot gain the tacit knowledge of what animal flesh looks and feels like from a computer."[48]

Opponents counter, however, that specimens do not resemble their living counterparts, because the processes and chemicals used to prepare the body for the dissection lab alter the physical structure of the organism. Formaldehyde used to preserve the body, for example, can misshape organs and can render body tissues gelatinous. As one doctor observed: "As a doctor who performs autopsies, I can assure students that computer images of well-preserved tissues look more like the 'real thing' than the squishy gray organs of a formalin-fixed specimen. Simulated dissection is very realistic, the accompanying text is elegant, and the graphics are superb."[49]

Balcombe is among those who believe that nonanimal alternatives teach physiological and biological concepts as well as animal dissection. In his book *The Use of Animals in Higher Education*, Balcombe sets forth his case for replacing learning methods that cause pain and death to animals with methods that do not. To this end, Balcombe gathered empirical data from a number of educational studies that were designed to measure the efficacy of nonanimal teaching methods. Balcombe cites close to 30 studies that conclude that students who used alternatives performed as well as or better than students using animals. Balcombe reports, too, that students in Sweden and Norway, where dissection is rarely used, scored higher in scientific literacy tests than their American counterparts. Balcombe concludes that "it is not known, nor is it easy to know, whether there is any relationship between the use of dissection as a teaching tool and the levels of scientific literacy of students who dissect. Hands-on learning methods are important and necessary, but they are abundantly available beyond dissected animals."[50]

> "Simulated dissection is very realistic, the accompanying text is elegant, and the graphics are superb."[49]
>
> — Nancy Harrison, MD.

Medical Training Alternatives

While many support alternatives to dissection in the precollege biology classroom, they feel there are valid arguments to morally justify the use of dissection at the university level. The common claim is that medical students must practice surgery and other medical techniques such as how to administer anesthesia on animals before

Ethics Courses

A growing number of universities today are requiring that students who may be called upon to experiment on animals—students in the biological sciences and medical or veterinary medicine, for example—take courses about the ethics of using research animals. These courses seek to make animal research more humane and also help students to identify and defend their own moral or religious ideologies in relation to animal research, an intensely personal issue for many students.

As Joy Mench, a professor of animal research who teachers an animal-research ethics course at the University of California–Davis, says: "We all realize now that it's a sensitive ethical issue, and people have to be free to be sensitive and have their concerns aired. . . . I get the whole range [of student reaction to animal research], from 'Absolutely unacceptable' to 'Why are we even talking about this?' Until we talk about it, sometimes people aren't really able to articulate why."

Whether these types of courses will help foster a more positive research culture and improve the welfare of laboratory animals will be seen in the years ahead.

Quoted in Josh Keller, "For Researchers on Animals, Ethics Training Is Sparse," *Chronicle of Higher Education*, September 19, 2008. https://chronicle.com.

they graduate to human patients. One researcher, Ted Valli, goes on to assert that "for students who plan to practice medicine, the more exposure they have to the sight, smell, and texture of tissues, the better their preparation to become confident clinicians."[51]

At the same time, a growing consensus appears to support the view that even medical training does not necessitate killing healthy animals; today the greatest move away from animal use in education has occurred in US medical schools. Just a generation ago, the terminal use of dogs and other animals was a routine component of medical school training—used to teach physiology, surgical techniques, and pharmacology. Changing attitudes about animal welfare, however, have led many top-ranked medical schools—including Columbia, Harvard, Stanford, and Yale—to join the trend away from using live animal methods. According to the Physicians Committee for Responsible Medicine, a group that opposes all forms of animal experimentation, more than 90 percent of medical schools in the United States no longer use live animal laboratories to train their medical students.

Many institutions rely instead on sophisticated versions of the alternatives used at lower educational levels—synthetic animal models and a wide variety of computer platforms, including surgical simulator tools. Vaughan Monamy describes one promising computer interface in his book *Animal Experimentation*:

Using a cadaver head, a neurosurgeon instructs two other doctors in a new brain surgery technique. Watching other surgeons work and practicing various medical techniques on cadavers are both considered useful and important teaching alternatives to animal experimentation.

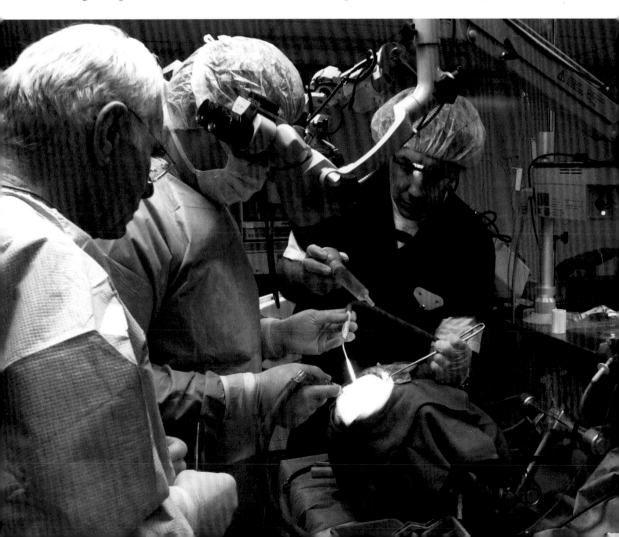

Virtual reality computer simulation offers users an opportunity to "perform" experiments such as laparoscopies without the need for a patient. By connecting a laparoscope to a virtual reality generator, images of what a surgeon "sees" as he or she "enters" a body are produced. Such technology is still in its infancy but currently holds the potential to outstrip other simulation methods in the near future.[52]

A growing number believe these and other alternatives render live-animal medical training obsolete.

The Value of Observation

At the same time, exposure to real surgery and trauma care remains a crucial part of medical training; clinical experiences and human-based methods can offer this type of experience. For example, apprentice teaching programs (apart from regular residency requirements) allow students to train in real medical settings, interacting with real patients, assisting experienced physicians and surgeons. Many schools, too, routinely send their students to observe and study a variety of surgical procedures in operating rooms and trauma centers. As the National Anti-Vivisection Society puts it, "Surgeons learn how to be surgeons from watching other surgeons."[53] Work on cadavers is another valuable resource for teaching physiology, surgical skills, and certain medical procedures, such as inserting a breathing tube into a patient's trachea.

The efficacy of alternatives at the highest levels of American education prompts many to question the necessity of using live animals in any type of research. As animal advocates Erin E. Williams and Margo Demello write in their book *Why Animals Matter: The Case for Animal Protection*:

The creation of alternatives to the use of live animals both in medical and veterinary schools does more than save the lives of those animals who would have been killed in classroom studies. Perhaps even more important, these alternatives can also substitute for the use of animals in much of biomedical

> "It is not known, nor is it easy to know, whether there is any relationship between the use of dissection as a teaching tool and the levels of scientific literacy of students who dissect."[50]
>
> — Author Jonathan Balcombe.

research. The question is, then, why is medical research so slow to follow in the footsteps of medical schools?[54]

Renewed and steady interest in the rights of animals has clearly affected their use in educational settings. As more realistic alternatives arise that can be used by all students in all grades, the question remains as to whether such alternatives will, or more importantly should, replace the use of animal subjects.

Facts

- Tufts University School of Medicine and other teaching institutions have implemented educational memorial programs that utilize donated cadavers for student dissection.

- According to the Humane Society of the United States, only about 10 of the nation's 125 accredited medical schools still use live animal labs to teach medical concepts.

- As reported by the Physicians Committee for Responsible Medicine, purchasing animal cadavers for dissection is expensive. Using nonanimal materials will result in significant economic savings to the educational establishment.

- The US Department of Agriculture reports that in 2007, over 72,000 dogs were used in research and education. Some of these were supplied by Class B dealers, which are licensed to buy dogs from random sources such as animal shelters or newspaper ads.

- According to one estimate by Jonathan Balcombe of the Humane Society, approximately 10 million to 12 million animals are killed for school dissection each year. Most of these are wild caught.

CHAPTER FIVE

What Are the Alternatives to Animal Experimentation?

The development of alternatives to animal-based research has progressed dramatically in recent years. To many the sheer volume of alternative scientific protocols available renders the continuation of research that causes animal suffering and death untenable. On the other hand, many believe that completely eliminating animals from the laboratory would compromise scientific progress and, specifically, impede the development of new treatments for the diseases that continue to plague human populations. Another viewpoint is that while some forms of animal research may be phased out as advanced technology becomes available, new scientific endeavors, such as the quest to fully understand human DNA, will necessitate the continued use of laboratory animals.

Despite the broad spectrum of views on the subject, most people agree that when animals are used in research, their suffering should be minimized whenever possible and that the development of alternatives to using animal models should be a paramount goal. Many scientific regulatory bodies have formulated guidelines to fulfill these tenets. The best known are a set of principles that serve as a basis for the design of any experiment using lab animals. Set forth by zoologist William Russell and microbiologist

Rex Burch in 1959, the "three Rs," refer to *reducing* the number of animals used in any experiment, *refining* experimental methods to minimize pain and suffering, and *replacing* experiments that use animals with other types of experiments whenever possible.

Today the field of research alternatives to reduce, refine, and replace animal models is vast. There are three primary alternatives to the use of lab animals: in vitro research, which typically involves cell cultures; in silico research, which utilizes computer-based and mathematical modeling systems; and studies that rely on human systems.

In Vitro Research

To date, many of the reductions in animal use have been due to the use of in vitro ("in glass") experiments. As opposed to in vivo ("in the living being") tests, in vitro tests are performed in test tubes, assay plates, or other types of containers. By creating artificial environments using human or animal cells or tissues, researchers gain

A medical researcher conducts testing on rabbits. Most people agree that pain and suffering should be minimized when animals are used in research and that finding alternatives to animal experimentation—where possible—is a worthy goal.

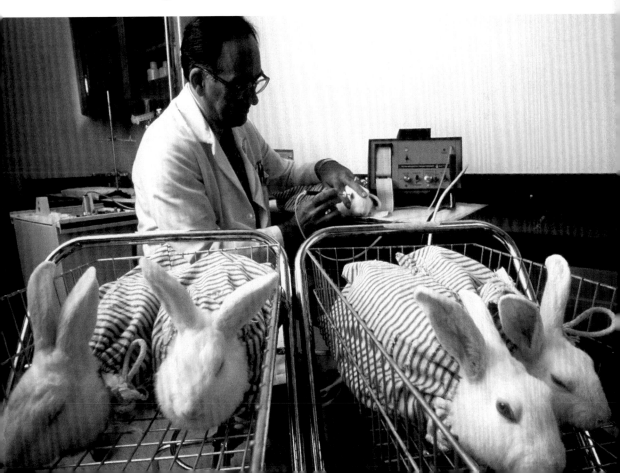

information about biological processes at the cellular level. One example of a successful in vitro application is the production of monoclonal antibodies, which are used in medical diagnostic tests and nearly every field of biomedical research today. Previously, the production of these antibodies involved the extremely painful process of injecting tumor cells into the abdominal cavities of mice. This process yielded a fluid rich in antibodies—and an excruciatingly distended abdomen for the mouse. The success of in vitro techniques for producing these antibodies has saved millions of rodents from the "antibody farms" of the past.

In vitro research is typically used to evaluate the effects and toxicity of certain drugs. Because many drug side effects involve the liver, researchers can administer carefully controlled doses to healthy liver cells—both human and animal—and observe negative effects. A recent innovation that could dramatically improve current in vitro drug screening is the development of two biochips, MetaChip and DataChip. Each computer chip contains human liver enzymes and cells and can be used to quickly determine whether particular chemicals or drugs are toxic. As Douglas Clark, a professor of chemical engineering at the University of California–Berkeley who is involved in the research, puts it: "Ensuring the safety of new compounds without testing them on animals presents a new challenge to the industry, especially as the number of compounds increases. These chips can meet this challenge by providing comprehensive toxicity data very quickly and cheaply."[55]

Another development that is making in vitro cultures more valuable as research tools involves growing cells within a three-dimensional structure. As opposed to single-layer cultures, this arrangement maintains organ-specific functions better, making them more predictive of human health effects. Today the Johns Hopkins University Center for Alternatives to Animal Testing is just one organization involved in the creation of these strikingly complex cell cultures that could transform toxicity testing. Erwin van Vliet, a scientist at Johns Hopkins, predicts that in the future "in vitro models will provide the main biological systems for toxicity testing instead of currently used whole-animal models."[56]

The emerging field of human stem cell technology has also radically altered the way scientists approach medical research.

Studies using stem cells—that is, cells with the ability to divide and differentiate into virtually any type of body cell—occur in vitro. Although this type of research is still in its infancy, scientists today can study how toxic chemicals affect these specialized cells, which may determine whether a particular substance imperils the development of a human fetus. As the field matures, many researchers believe that it could replace the type of animal research that results in fetal deformities and death. Other promising research with stem cells explores the possibility of programming stem cells to replace damaged cells in patients suffering from diabetes, arthritis, Parkinson's disease, and other illnesses.

Despite these scientific breakthroughs, many believe that in vitro research is destined to remain limited in scope. The primary criticism is that cell cultures and other in vitro techniques do not allow researchers to evaluate whole body experiences in a living organism, such as changes in blood pressure or immunological function, or idiosyncratic effects such as routes of dosage or long-term exposure. Simply put, in vitro techniques cannot accurately replicate the biological systems found in even very simple life forms. For these reasons, many researchers—even those who agree that animal models are sometimes poor predictors of human results—still see a role for laboratory animals.

"[In the future] in vitro models will provide the main biological systems for toxicity testing instead of currently used whole-animal models."[56]

— Erwin van Vliet, a researcher at the Johns Hopkins Center for Alternatives to Animal Testing.

In Silico Research

In silico research refers to the use of computer-based modeling and mathematical systems that simulate chemical, molecular, and cellular processes to predict the likely biological effects on a living being. As the technology increases, computer-based research has the potential to replace millions of lab animals each year.

For example, computers can virtually screen millions of chemical combinations and theoretically remove the need for testing drug toxicity on a live animal model. The selected computer-aided drug compounds can be further tested on computer models of patients. One type of research at the National Cancer Institute in the 1980s utilized mathematical modeling to analyze the immune system's response to cancer. Whereas it was previously believed that

Cancer Epidemiology

Many thousands of animals are employed in cancer research each year in the United States. The epidemiological study is one alternative to gathering lifesaving information about the incidence and cause of various types of cancers by using laboratory animals. Epidemiological studies highlight not only which groups get certain cancers, but also which groups do not. For example, colon cancer is a leading cause of death in more industrialized countries such as Australia, Canada, the United Kingdom, the United States, and those in western Europe. By comparison, Japan and Nigeria have relatively low rates of this type of cancer.

Epidemiological studies revealed that second-generation migrants from Japan, a low-risk country, take on the same risk of developing colon cancer as other Americans after living in the United States. This observation led to the understanding that colon cancer is associated with certain environmental factors—including a diet high in fat, red meat, and alcohol, as is typical in the United States. Many animal rights advocates argue that this type of research is more useful than animal testing. It has yielded more valuable data and has contributed more to lowering cancer rates by identifying how particular types of cancer are acquired.

the immune system always suppressed cancer growth, this research confirmed that the immune system could actually *stimulate* cancer growth. As biophysicist Charles DeLisi, one of the lead researchers, commented: "If our model had been around ten years ago, it could have predicted what it's taken scientists countless man-hours and animals to figure out. This is the value of mathematical modeling—it comes up with things that you might otherwise miss."[57]

In another application, in 2007, researchers developed an in silico model of tuberculosis that has boosted understanding of the disease's progression and has been instrumental in the discovery of pharmaceutical treatments. One of the prime benefits to this type of research is that it is faster than real time, allowing relevant effects to be observed in minutes rather than months.

In addition, computers can quickly and efficiently manage and mine prodigious amounts of data, including medical records and genetic information. This enables researchers to see patterns or make vital scientific connections about the progression and treatment of disease. Perhaps most importantly, research institutes can maintain databases on past and ongoing research. Making this data widely accessible will ensure that research on animals is not needlessly repeated. It also allows researchers to share information about developing less painful techniques when animal models are necessary. One way scientists share their knowledge and expertise is through a worldwide Internet forum for lab animal researchers. As one user describes it: "More than 2000 veterinarians, scientists, technicians, and others vibrantly exchange information and experiences in fine-tuning and refining animal experiments. If I want expert advice and anecdotal experiences on the best painkillers for a particular type of monkey surgery, I post my question and within a day receive a wealth of information."[58]

Despite the demonstrated benefits, there remains one overriding drawback to in silico research: Even though scientific knowledge of biological functions has burgeoned in recent years, researchers still have only a limited understanding of how the body works, especially on a cellular level. Because this information cannot be programmed into a computer, the results may not be as predictive as research on real organisms. As one researcher describes the problem:

> To model mathematically a whole animal or animal system in a completely adequate way, so that experiments with the model would yield exactly the same results as experiments with the living animal (and hence be able to replace it), we would need to know everything there is to know about the animal or animal system in question and would no longer need to do research on either animals or alternatives![59]

"If our [mathematical] model had been around ten years ago, it could have predicted what it's taken scientists countless man-hours and animals to figure out."[57]

— Biomedical researcher Charles DeLisi.

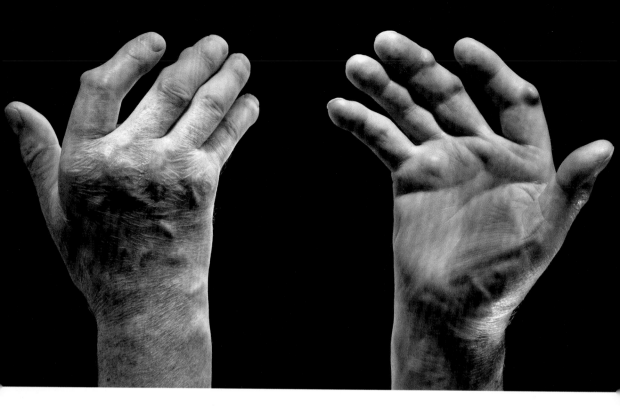

Human stem cell research, still an emerging field, might eventually replace some types of animal experimentation. Stem cell research offers promise for replacing damaged cells in people suffering from various diseases, including arthritis. Pictured are hands deformed by arthritis.

According to the National Institutes of Health, animal research and computer modeling are interconnected and are likely to remain so for years to come: "Data for computer models often comes from animal studies. In turn, computer models reveal gaps for further studies in living organisms."[60]

Human Studies

In addition to the ethical dimension to most arguments to reduce animal experimentation, many opponents cite practical justifications, mainly that animal models are too ambiguous and that the results cannot always be extrapolated to human populations. In their book *Sacred Cows and Golden Geese*, C. Ray Greek and Jean Swingle Greek describe why they believe it is imperative to approach biomedical research in a purely human context. At the same time, they address some of the moral arguments put forth by those who condone animal research:

> There is a ludicrous scare tactic perpetrated by animal experimenters and their lobbyists. That is the claim that if there were no animal experiments we would have to exper-

iment on humans. Human experiments, yes, but not on caged humans, nor prisoners, nor the mentally disabled, nor lab humans, nor any unwilling experimental humans. We would conduct experiments on human cells and human tissues, examine and document humans at autopsy, tally and analyze the results of human epidemiology studies, more carefully observe humans in the clinical setting and spread the word among humans on preventative measures. It is human health that is at risk and human wellness that is our objective. Is it not reasonable to observe the species that needs curing directly?[61]

The epidemiological study—that is, the study of a disease that afflicts an entire population—is one important tool in understanding the genesis and prevention of human diseases. For example, after World War II, the observation that poorly fed Europeans had lower cholesterol than Americans with higher-fat diets led to studies that confirmed what is now generally accepted: that a diet high in fat can lead to heart disease. Other epidemiological studies have provided useful medical insight in other areas—the effects of smoking on lung cancer, for example—and may be bolstering the move away from experiments that rely on animal data.

> "Data for computer models often comes from animal studies. In turn, computer models reveal gaps for further studies in living organisms."[60]
>
> — National Institutes of Health.

Clinical trials—in which new drugs and other treatments are administered to human volunteers—have also aided medical discovery for years. Generally, a new drug will be tested on cells in a laboratory before it is tested on animals. If it appears safe, it will be administered to a limited number of human volunteers, often patients with specific health conditions. To date, the clinical study remains the best way to observe what a drug will do in a human recipient. In recent years a process called microdosing has been increasingly utilized. With microdosing, healthy human volunteers are given a tiny dose of a drug. There is virtually no risk to the human host, because the dose, while high enough to produce observable effects at the cellular levels, is far below that which would produce dangerous, toxic side effects. Traditionally, this type of testing has been performed on animals. Many hope that these newer techniques could negate the need for the large volume of animals currently enlisted—that is, that trials could move from the in vitro testing of cells directly to human subjects, bypassing the animal trial stage altogether.

Noninvasive diagnostic imaging technology is another sophisticated alternative to the study of animal subjects. State-of-the-art imaging technology that allows researchers to peer into the human body and observe functioning biological systems include ultrasound, magnetic resonance imaging (MRI), positron-emission tomography (PET), and computerized axial tomography (CAT). In 2006, Stanford researchers demonstrated a new technique called "fly-through" that uses software to produce CAT and PET images rendered in three dimensions, giving researchers a startlingly clear picture of a person's interior.

Human autopsies are another valuable source of data. As the Greeks put it:

> If you want to know what caused a failure, investigate the failed entity. . . . Research in diabetes, hepatitis, appendicitis, rheumatic fever, typhoid fever, ulcerative colitis, congenital heart disease, hyperparathyroidism, and many other illnesses has been enriched by autopsies. Autopsies elucidated the mechanisms of shaken baby syndrome, sudden infant death syndrome, and head injuries suffered during car accidents.[62]

Today the Visible Human Project at the National Library of Medicine is utilizing data from these sources to provide unique views of the body. The project has produced anatomically complete, three-dimensional representations of human bodies using human cadaver cross sections and electronic images from MRI and CAT scans. These high-tech representations aid biomedical research and also provide training in surgical and other medical techniques.

A Change in Thinking May Be on the Way

As scientists continue to plumb the mechanisms of human disease, these and forthcoming alternatives will be put to the test.

"It is human health that is at risk and human wellness that is our objective. Is it not reasonable to observe the species that needs curing directly?"[61]

— Medical doctor C. Ray Greek and veterinarian Jean Swingle Greek.

Alternatives in Safety Testing

The drive to reduce, refine, and replace animal testing extends far beyond the biomedical research community. Today a number of companies have renounced using animals to test consumer products such as food additives, skin-care products, industrial chemicals, and many other products. The National Institute of Environmental Health Sciences and over a dozen other federal agencies, collectively known as the Interagency Coordinating Committee on the Validation of Alternative Methods, have been at the forefront of this movement. This interagency effort has developed a process to identify new test methods that do not rely on animal models. As opposed to the notorious Draize tests of the past, in which caustic products were applied directly to the eyes and skin of live animals, companies today might use reconstructed human skin models to test for skin irritation. In vitro methods are commonly employed to test whether a particular chemical or product will harm the eye. Because of these and other alternatives, the number of animals required for safety tests has been dramatically reduced.

Whether they will unequivocally fulfill the tenet to reduce, refine, and replace animal experimentation is uncertain. What is certain is that any move from a system based on animal testing to one that focuses primarily on nonanimal models will likely require a change in thinking. According to Johns Hopkins's van Vliet: "Toxicology has started to give up on traditional animal-based testing and is moving towards the implementation of new technologies expected to provide a more accurate, science-driven, and humane assessment of human toxicological risk."[63] Whether this change in thinking will, or should, extend to all forms of animal-based research will be determined in the years ahead.

Facts

- In 2008 the National Institutes of Health and the Environmental Protection Agency began a five-year collaborative research project to reduce the number of animals used in toxicity testing. The program will utilize robotic technology and in vitro techniques instead of laboratory animals.

- Today genetically modified animals are used as models in many areas of biomedical research. Transgenic mice, for example, contain additional foreign DNA in every cell and are used to study gene function and a variety of diseases.

- Bioengineers at Brown University are now able to develop complex, three-dimensional cellular structures that could eventually lead to models that replicate human organs, which could partially reduce the need for animal-based research.

- Xenotransplantation is an area of biomedical research that involves transplanting tissues or organs from one species to another as a way of addressing the shortage of human organs needed for transplants.

- Current research at the Massachusetts Institute of Technology involves placing a computer chip in human liver tissue, which gives researchers a way to study how the liver metabolizes drugs and other chemicals without the use of animal models.

- Nanotechnology—manipulating matter on an atomic and molecular scale—has vast implications in biomedical research. The Japanese company Transparent, for example, has devised a way to apply a nanoengineered coating to liver tissue cultures, which makes them easier to study on a molecular level.

Related Organizations and Websites

American Anti-Vivisection Society (AAVS)
801 Old York Rd., Suite 204
Jenkintown, PA 19046
phone: (800) 729-2287
website: www.aavs.org

Established in 1883, the AAVS is the oldest and one of the most prominent animal protection groups in the United States. The society opposes all forms of animal experimentation and sponsors research on alternatives to using live animal models.

American Association for Laboratory Animal Science (AALAS)
9190 Crestwyn Hills Dr.
Memphis, TN 38125
phone: (901) 754-8620
fax: (901) 759-5849
website: www.aalas.org

The AALAS is a professional association of animal care workers, veterinarians, and manufacturers involved in the production, care, and study of animals used in biomedical research, which the organization believes is essential for medical progress. The AALAS produces a variety of educational DVDs and other publications.

Americans for Medical Progress (AMP)

908 King St., Suite 301
Alexandria, VA 22314
phone: (703) 836-9595
fax: (703) 836-9594
website: www.amprogress.org

AMP supports scientists' humane use of animals in research for the identification, study, and treatment of disease and injury. The group's website catalogs many medical advances that involved animal research, including the treatment of asthma and HIV and the development of certain vaccines and antibiotics.

Animal Liberation Front (ALF)

21044 Sherman Way, No. 211
Canoga Park, CA 91302
phone: (818) 932-9997
fax: (818) 932-9998
e-mail: press@animalliberationpressoffice.org
website: www.animalliberationfront.com

ALF seeks to end animal exploitation and reduce the suffering of animals throughout the world. The organization encourages activism on behalf of all animals; its website provides stories of animal activism, news articles, profiles of activists, and a variety of educational materials.

Animal Welfare Institute (AWI)

900 Pennsylvania Ave. SE
Washington, DC 20003
phone: (202) 337-2332
fax: (202) 446-2131
e-mail: awi@awionline.org
website: www.awionline.org

The AWI was founded in 1951 to alleviate suffering inflicted on animals by people, with a strong emphasis on animals used for biomedical experimentation, which the group opposes. The institute publishes numerous books, reports, brochures, and a quarterly magazine.

Foundation for Biomedical Research (FBR)

818 Connecticut Ave. NW, Suite 900
Washington, DC 20006
phone: (202) 457-0654
fax: (202) 457-0659
e-mail: info@fbresearch.org
website: www.fbresearch.org

The FBR is a nonprofit educational organization established in 1981 to inform the American public about the proper and necessary role of animal models in biomedical research and toxicity testing. The foundation promotes public support for animal research through a variety of educational programs and publishes the magazine *ResearchSaves*.

Fund for Animals

200 W. Fifty-Seventh St.
New York, NY 10019
phone: (888) 405-3863
website: www.fundforanimals.org

The Fund for Animals was founded by prominent animal advocate Cleveland Amory in 1967 to promote animal welfare throughout the world. The fund opposes all forms of animal experimentation and seeks to protect animals in the courts and through the provision of sanctuary at animal-care facilities.

In Defense of Animals

3010 Kerner Blvd.
San Rafael, CA 94901
phone: (415) 448-0048
fax: (415) 454-1031
e-mail: idainfor@idausa.org
website: www.idausa.org

In Defense of Animals is a nonprofit organization founded in 1983 to protect the rights, welfare, and habitat of animals throughout the world. The organization seeks to abolish the exploitation and abuse of animals in the nation's biomedical laboratories through the publication of opinion pieces, fact sheets, and brochures that highlight animal abuse in the name of science.

Institute for In Vitro Sciences (IIVS)
30 W. Watkins Mill Rd., Suite 100
Gaithersburg, MD 20878
phone: (301) 947-6523
fax: (301) 947-6538
website: www.iivs.org

Founded in 1997, the IIVS is a nonprofit, science-based organization dedicated to advancing alternatives to animal-based research methods. To this end, the institute works with government agencies, industry leaders, and medical laboratories to implement in vitro and other testing strategies that limit animal use while furthering medical and scientific advances.

National Institutes of Health
9000 Rockville Pike
Bethesda, MD 20892
phone: (301) 496-5793
website: www.nih.gov

The National Institutes of Health is part of the US Department of Health and Human Services. In addition to its own research, the institute supports animal-based experimentation through the disbursement of funds to the nation's universities and research institutes. It also sets policy on the welfare of laboratory animals.

People for the Ethical Treatment of Animals (PETA)
501 Front St.
Norfolk, VA 23510
phone: (757) 622-7382
fax: (757) 622-0457
website: www.peta.org

PETA is an international animal rights organization with more than 2 million members and participants worldwide. As one of the most prominent groups to speak out against the inhumane use of animals, PETA implements a variety of programs, including public education, cruelty investigations, animal rescue, and celebrity advertisements that promote the rights of animals.

Physicians Committee for Responsible Medicine (PCRM)

5100 Wisconsin Ave. NW, Suite 400
Washington, DC 20016
phone: (202) 686-2210
e-mail: pcrm@pcrm.org
website: www.pcrm.org

The PCRM was founded in 1985 as a network of physicians and laypersons to promote higher standards for ethics and effectiveness in medical research. The group promotes alternatives to animal research and nonanimal methods in medical education. It publishes the magazine *Good Medicine* and numerous fact sheets about animal-based research.

Additional Reading

Books

Tom L. Beauchamp et al., *The Human Use of Animals: Case Studies in Ethical Choice*. New York: Oxford University Press, 2008.

Christina Campbell et al., *Primates in Perspective*. New York: Oxford University Press, 2010.

P. Michael Conn and James V. Parker, *The Animal Research War*. New York: Palgrave Macmillan, 2008.

C. Ray Greek and Jean Swingle Greek, *What Will We Do If We Don't Experiment on Animals?* Victoria, BC: Trafford, 2006.

The Institute for Laboratory Animal Research, *Guide for the Care and Use of Laboratory Animals*. Washington, DC: National Academies Press, 2010. (Full-text can be viewed online at: www.nap.edu/catalog.php?record_id=12910#toc.)

Vaughan Monamy, *Animal Experimentation: A Guide to the Issues*. Cambridge: Cambridge University Press, 2009.

Peter Singer, *Animal Liberation: The Definitive Classic of the Animal Rights Movement*. New York: HarperCollins, 2009.

Paul Waldau, *Animal Rights: What Everyone Needs to Know*. New York: Oxford University Press 2011.

Periodicals

Eryn Brown, "Poll: Scientists Say Animal Research Ethically Complicated, but Necessary," *Los Angeles Times*, February 23, 2011.

Marcia Clemmitt, "Animal Rights," *CQ Researcher*, January 8, 2010.

Meredith Cohn, "Alternatives to Animal Testing Gaining Ground," *Baltimore Sun*, August 26, 2010.

P. Michael Conn and James V. Parker, "Winners and Losers in the Animal Research War," *American Scientist*, May/June 2008.

Thomas Hartung, "Toxicology for the Twenty-First Century," *Nature*, July 2009.

Anna Wilde Mathews, "Recent Cases Point to the Limitations of Animal Drug Tests," *Wall Street Journal*, March 30, 2007.

Robin McKie, "Ban on Primate Research Would Be Devastating, Scientists Warn," *Observer* (London), November 2, 2008.

Richard Monastersky, "Protestors Fail to Slow Animal Research," *Chronicle of Higher Education*, April 18, 2008.

Katherine Perlo, "Would You Let Your Child Die Rather than Experiment on Animals? A Comparative Questions Approach," *Society and Animals*, 2003.

Andrew Read, "Vivisectionists Strike Back," *Nature*, May 29, 2008.

Peter Singer and Richard A. Posner, "Animal Rights," *Slate*, June 15, 2009.

Wesley Smith, "Human Guinea Pigs," *Weekly Standard*, January 3, 2006.

Valerie Strauss, "When Cutting Up in Class Is Okay," *Washington Post*, March 5, 2007.

Meredith Wadman, "Medical Schools Swap Pigs for Plastic," *Nature*, May 8, 2008.

Source Notes

Introduction: Species and the Role of Animal Experimentation

1. Quoted in C. Ray Greek and Jean Swingle Greek, *Sacred Cows and Golden Geese: The Human Cost of Experiments on Animals*. New York: Continuum, 2000, p. 10.

2. Donna J. Haraway, *When Species Meet*. Minneapolis: University of Minnesota Press, 2008, p. 69.

3. Haraway, *When Species Meet*, p. 70.

Chapter One: What Are the Origins of the Animal Experimentation Debate?

4. Jeremy Bentham, *The Works of Jeremy Bentham*, vol. 1: *Principles of Morals and Legislation, Fragment on Government, Civil Code, Penal Law*, 1843. Reprinted on The Online Library of Liberty, 2011. http://oll.libertyfund.org.

5. Quoted in *Veterinary Heritage*, "A History of Antivivisection from the 1800s to the Present: Part I (mid-1800s to 1914), May 2008. Republished in Black Ewe. http://brebisnoire. wordpress.com.

6. Quoted in *Veterinary Heritage*, "A History of Antivivisection from the 1800s to the Present."

7. Charles Darwin, *The Life and Letters of Charles Darwin*, vol. 3. London: John Murray, 1887, p. 660.

8. Henry Salt, "*Animals' Rights*: The Principles of Animals' Rights," Animal Rights History, 2003. www.animalrightshistory.org.

9. Quoted in *Veterinary Heritage*, "A History of Antivivisection from the 1800s to the Present."

10. Stop Animal Exploitation Now!, "Animal Experimentation in the United States," All-Creatures.org, 2007. www.all-creatures. org.

11. Larry Carbone, *What Animals Want: Expertise and Advocacy in Laboratory Animal Welfare Policy*. New York: Oxford University Press, 2004, p. 239.

Chapter Two: Is Animal Research Necessary for Medical Progress?

12. Quoted in William Crawley, "Peter Singer Defends Animal Experimentation," BBC, November 26, 2006. www.bbc.co.uk.

13. Quoted in Joanne Zurlo, Deborah Rudacille, and Alan M. Goldberg, "Animals and Alternatives in Testing, History, Science, and Ethics," Johns Hopkins University Center for Alternatives to Animal Testing, 1994. http://caat.jhsph.edu.

14. Carl Cohen and Tom Regan, *The Animal Rights Debate*. Oxford, England: Rowman and Littlefield, 2001, p. 71.

15. Cohen and Regan, *The Animal Rights Debate*, pp. 11–12.

16. Quoted in Jennifer Brown, "About 45,000 People Are Diagnosed with Head and Neck Cancer Each Year in the United States. About 15,000 Die from the Disease. Cancerous-Tumor Research Riding on the Backs of Mice," *Denver Post*, June 17, 2010. www.denverpost.com.

17. Charles Alden, "In Memory of a Million Mice," *World and I*, May 2007. www.worldandi.com.

18. Gary Wolf, "Of Mice and Men," *Wired*, March 2010, p. 82.

19. Greek and Greek, *Sacred Cows and Golden Geese*, p. 151.

20. Quoted in Greek and Greek, *Sacred Cows and Golden Geese*, p. 139.

21. Greek and Greek, *Sacred Cows and Golden Geese*, p. 137.

22. Quoted in Greek and Greek, *Sacred Cows and Golden Geese*, p. 153.

23. Quoted in Science Daily, "Mouse Grimace Scale to Help Identify Pain in Humans and Animals," May 10, 2010. www.sciencedaily.com.

24. PETA, "Hasn't Every Major Medical Advance Been Attributable to Experiments on Animals?," 2011. www.peta.org.

25. PETA, "Xenografts: Frankenstein Science," 2011. www.peta.org.

Chapter Three: Should Primates Be Used in Research?

26. Quoted in Peter Singer, ed., *In Defense of Animals*. New York: Basil Blackwell, 1985, p. 138.

27. Wesley J. Smith, "A Monkey for Your Grandmother: Animal-Liberationists Force Medical Research on the Backburner," *National Review Online*, February 10, 2004. www.discovery.org.

28. California Biomedical Research Association, "Fact Sheet: Primates in Biomedical Research," January 11, 2011. www.cabiomed.org.

29. Frederick King et al., "Primates," *Science*, June 10, 1988. www.sciencemag.org.

30. Quoted in Robin McKie, "Ban on Primate Experiments Would Be Devastating, Scientists Warn," *Observer* (London), November 2, 2008. www.theobserver.com.

31. Jarrod Bailey, "Non-human Primates in Medical Research and Drug Development: A Critical Review," *Biogenic Amines*, July 2005. www.pcrm.org.

32. Bailey, "Non-Human Primates in Medical Research and Drug Development."

33. Greek and Greek, *Sacred Cows and Golden Geese*, p. 195.

34. Bailey, "Non-Human Primates in Medical Research and Drug Development."

35. California National Primate Research Center, "In the News," November 2010. www.cnprc.ucdavis.edu.

36. American Society of Primatologists, "Research Questions and Answers," 2011. www.asp.org.

37. Great Ape Project, "GAP Project," 2011. www.greatapeproject.org.

38. Quoted in Great Ape Project, "Mission and Vision," 2011. www.greatapeproject.org.

Chapter Four: Animals in Education

39. Jonathan Balcombe, *The Use of Animals in Higher Education: Problems, Alternatives & Recommendations*. Washington, DC: Humane Society, 2000, p. 27.

40. Tom Regan, *Empty Cages: Facing the Challenge of Animal Rights*. Lanham, MD: Rowman and Littlefield, 2004, p. 161.

41. Susan Offner, "The Importance of Dissection in Biology Teaching," *American Biology Teacher*, March 1993, p. 147.

42. Linda Birke, Arnold Arluke, Mike Michael, *The Sacrifice: How Scientific Experiments Transform Animals and People*. West Lafayette, IN: Purdue University Press, 2007, p. 84.

43. Birke et al., *The Sacrifice*, p. 87.

44. Quoted in Balcombe, *The Use of Animals in Higher Education*, p. vii.

45. Quoted in Balcombe, *The Use of Animals in Higher Education*, p. 16.

46. Quoted in Birke et al., *The Sacrifice*, p. 84.

47. George K. Russell, *Laboratory Investigations in Human Physiology*. New York: Macmillan, 1978, p. vi.

48. Birke et al., *The Sacrifice*, p. 89.

49. Nancy L. Harrison, "A Doctor's View of Dissection," Dissection Alternatives, 2001. www.dissectionalternatives.org.

50. Balcombe, *The Use of Animals in Higher Education*, p. 8.

51. Ted Valli, "Dissection: The Scientific Case for a Sound Medical Education," *Journal of Applied Animal Welfare Science*, 2001, p. 129.

52. Vaughan Monamy, *Animal Experimentation: A Guide to the Issues*. Cambridge: Cambridge University Press, 2009, p. 82.

53. National Anti-Vivisection Society, "If Medical Students Couldn't Practice Surgical Techniques on Animals, Wouldn't That Compromise the Learning Process and Put Their Future Patients at Risk?," 2004. www.navs.org.

54. Margo Demello and Erin E. Williams, *Why Animals Matter: The Case for Animal Protection.* New York: Prometheus, 2007, p. 210.

Chapter 5: What Are the Alternatives to Animal Experimentation?

55. Quoted in University of California, "New Biochip Could Replace Animal Testing," December 18, 2007. www.university ofcalifornia.edu.

56. Erwin van Vliet, "Current Standing and Future Prospects for the Technologies Proposed to Transform Toxicity Testing in the 21st Century," ALTEX 28, Johns Hopkins University, January 2011. altweb.jhsph.edu.

57. Quoted in Natalie Angier, "The Electronic Guinea Pig," *Discover*, September 1983, p. 8.

58. Carbone, *What Animals Want*, p. 242.

59. Quoted in Michael Allen Fox, *The Case for Animal Experimentation: An Evolutionary and Ethical Perspective.* Berkeley and Los Angeles: University of California Press, 1986, p. 178.

60. National Institutes of Health, "Medical Research with Animals," January 19, 2011. http://nih.gov.

61. Greek and Greek, *Sacred Cows and Golden Geese*, p. 99.

62. Greek and Greek, *Sacred Cows and Golden Geese*, p. 104.

63. Van Vliet, "Current Standing and Future Prospects for the Technologies Proposed to Transform Toxicity Testing in the 21st Century."

Index

Note: Boldface page numbers indicate illustrations.

Alden, Charles, 29
alternatives to animal
   experimentation
   human studies
      autopsy/cadaver studies, **65**, 76
      clinical trials, 30, 75–76
      epidemiological studies, 75
      noninvasive diagnostic imaging
         technology, 76
      stem cell research, 74
   in silico research, 71–74
   in vitro research, 69–71
   in medical schools, 63–67
   in secondary and undergraduate
      education, 62–66
American Anti-Vivisection Society
   (AAVS), 16, 79
American Association for
   Accreditation of Laboratory
   Animal Care, 21
American Association for Laboratory
   Animal Science (AALAS), 79
Americans for Medical Progress
   (AMP), 80
American Society for the Prevention
   of Cruelty to Animals (ASPCA),
   16
American Society of Primatologists,
   54
anesthetics/analgesics, 23
Animal Enterprise Terrorism Act
   (AETA, 2006), 14, 54
Animal Experimentation (Monamy),
   65–66
Animal League Defense Fund, 18
Animal Liberation (Singer), 18, 24

Animal Liberation Front (ALF), 18,
   80
animal procurement industry, 56
animal protection/animal rights
   movement
   early influences on, 13–15
   in modern times, 18–20
   origins of, 11
      in US, 15–17
animal research
   arguments in defense of, 6–7,
      13, 26–29, 37–38, 44–45, 48,
      58–59
   arguments opposing, 14, 15–17,
      31–32, 38, 43–44, 48–49,
      59–60
   Hall's principles for, 25–26
   See also alternatives to animal
      experimentation
animal shelters, 16
Animal Welfare Act (AWA, 1966),
   18–19
   animals covered/excluded under,
      20, 23
   limitations of, 22
Animal Welfare Institute (AWI), 80
autopsies, human, 76
Aziz, Tipu, 24

Bailey, Jarrod, 48, 49
Balcombe, Jonathan, 57, 63, 66, 67
Beers, Diane, 19
behavioral research, animals used in,
   23
Bell, Charles, 11–12
Bentham, Jeremy, 10
Bernard, Claude, 12, **13,** 15
Biogenic Amines (journal), 48
Birke, Linda, 58–59

Black Beauty (Sewell), 14–15
Boss, Irwin, 31–32, 36
Brown University, 78
Burch, Rex, 69

California Biomedical Research
   Association, 45
California National Primate Research
   Center, 49
cancer epidemiology, 72
Carbone, Larry, 22
cats, 57
   numbers used in research, 22
Cavalieri, Paola, 51
chimpanzees, genetic similarities
   between humans and, 45
class A (animal) dealers, 40
class B (animal) dealers, 67
clinical trials, 30, 75–76
Cohen, Carl, 26–27
computer-based modeling, 71
computerized axial tomography
   (CAT), 76
constraint-induced movement
   therapy, 44–45
Convention on International Trade
   in Endangered Species (CITES),
   54
Coordinating Committee on the
   Validation of Alternative Methods,
   77
Council on Defense of Medical
   Research, 17
Cruelty to Animals Act (UK, 1876),
   11

Darwin, Charles, 13, 14
DeLisi, Charles, 72, 73
Demello, Margo, 66
Department of Agriculture, US
   (USDA), 20, 40, 67
Descent of Man, The (Darwin), 13
dissection, in classroom, 55–56
   alternatives to, 62–66
   collection of animals for, 56–58
   number of animals killed annually
      for, 67

psychological impact on students,
   59–60
value of, 58–59
dogs
   numbers used in research, 22, 67
   in nutritional research, 36, **37**
   use by medical schools, 60, 64–65
   use in space program, 18, **21**
"A Dog's Tale" (Twain), 17
drug research, 32–33, 40
   clinical trials in, 75–76

Elvehjem, Conrad, 36
Empty Cages (Regan), 57
Environmental Protection Agency
   (EPA), 78
epidemiological studies, 75
   on cancer, 72
ethics courses, for students in
   biological sciences, 64

Food and Drug Administration, US
   (FDA), 37
For the Prevention of Cruelty (Beers),
   19
Fossey, Dian, 41–42
Foundation for Biomedical Research
   (FBR), 81
Freischlag, Julie, 61
Frolich, Theodor, 35
Fund for Animals, 81

gene therapy, in xenotransplantation,
   39
Goodall, Jane, 6–7, 8, 41–42, 46, 60
Graham, Jennifer, 55
Great Ape Project (GAP), 51, 53
Greek, C. Ray, 29–30, 32, 34
   on animal experimentation on
      HIV, 49, 53
   on failure of animal
      experimentation, 31
   on scare tactics used by animal
      experimenters, 74–75
Greek, Jean Swingle, 30, 32, 34
   on animal experimentation on
      HIV, 49, 53

on failure of animal
experimentation, 31
on scare tactics used by animal
experimenters, 74–75
*Guide for the Use and Care of
Laboratory Animals* (National
Academy of Sciences), 21, 23
on primates, 52

Hall, Marshall, 25
Haraway, Donna J., 8
Harlow, Harry, 46
Harrison, Nancy, 63
HIV (human immunodeficiency
virus), 39, 53
advances in research without using
animal models, 49
Holst, Axel, 35
Humane Society of the United
States, 67

imaging technology, diagnostic, 76
In Defense of Animals, 81
in silico research, 71–74
Institute for In Vitro Sciences (IIVS),
82
Institutional Animal Care and Use
Committee (IACUC), 20, 21
in vitro research, 69–71

Jimeno, Antonio, 27, 32
Johns Hopkins Center for
Alternatives to Animal Testing, 40

King, Frederick, 45
Koch, Robert, 16
Koko (mountain gorilla), 41

Laika (Russian space dog), 18, **21**
Lewontin, Richard C., 30
Lister, Joseph, 16

Magendie, François, 11–12
Massachusetts Institute of
Technology (MIT), 78
mathematical modeling, 71–72, 73
Mayock, Dennis, 61

McCollum, Elmer, 36
medical schools, use of live animals
in, 61
alternatives to, 63–67
Mellanby, Edward, 36
Mench, Joy, 64
mice/rats, **30**, 40
in cancer research, 27
dissection of, **56**
expression of discomfort by, 23
in pain research, 34
as research animals of choice,
22–23
transgenic, 78
Mitchell, S. Weir, 16
Mogil, Jeffrey, 34
Monamy, Vaughan, 65–66
monkeys, **25, 50**
maternal separation experiments
on, 46
rhesus macaque, **7**
in brain research, 42–44
Morgan, Thomas Hunt, 33
Morris, Roger, 48
mountain gorillas, 41–42, **42**
Mouse Grimace Scale, 34
Murray, Joseph, 33

nanotechnology, 78
National Academy of Sciences, 21
National Anti-Vivisection Society, 66
National Cancer Institute, 71
National Institute of Environmental
Health Sciences, 77
National Institutes of Health (NIH),
18, 75, 78, 82
National Review (magazine), 44
Nature (journal), 40
Nobel Prize, 33
nutritional studies, 35–36

Offner, Susan, 58
opinion polls. *See* surveys
organ transplantation
experiments on techniques for, 33
See also xenotransplantation
Origin of Species, The (Darwin), 13

Pacheco, Alex, 43–44
pain research, 33–34
Parkinson's disease research, 24
  using nonhuman primates, 48
pellagra, 35, 36
People for the Ethical Treatment of
  Animals (PETA), 18, 35, 38,
  82
Physicians Committee for
  Responsible Medicine (PCRM),
  67, 83
Policy on Humane Care and Use of
  Laboratory Animals (US Public
  Health Service), 20–21
polls. *See* surveys
positron emission tomography
  (PET), 76
primates, nonhuman, **42**, **50**, 54
  criticism of use in research, 48–49
  guidelines for care of, 52
  numbers used in research, 22,
    46–47
  protections for, 51, 52
  recent research using, 47–48
  similarities between humans and,
    45–46
  in spinal cord injury research,
    49–51
Public Health Service, US, 20–21

rats. *See* mice/rats
Regan, Tom, 57
research animals
  decline in numbers used, 22
  guidelines on care of, 17, 26
  numbers used in, 40
rhesus macaque monkeys, 7
  in brain research, 42–44
rickets, 35, 36, **37**
Rockefeller Institute for Medical
  Research, 17–18
Rose, Steven, 28
Rothschild, Miriam, 60
Royal Society for the Prevention of
  Cruelty to Animals (RSPCA), 11
Russell, William, 68–69

Sacred Cows and Golden Geese (Greek
  and Greek), 29–30, 74–75
safety testing, 19
  alternatives to animal models in,
    77
Salt, Henry, 15
scurvy, 35
Sewell, Anna, 14–15
Silver Spring monkey case, 42–44
Singer, Peter, 18, 24, 51
Smith, Wesley J., 44–45
Society for the Prevention of Cruelty
  to Animals (SPCA), 11
Starzl, Thomas, 38
stem cell research, 74
surveys
  of biomedical students on need for
    animal research, 40
  of scientists versus general public
    on support of animal testing, 9

Taub, Edward, 42, 44
Thomas, E. Donnall, 33
Twain, Mark, 17

*Use of Animals in Higher Education,
  The* (Balcombe), 63

Valli, Ted, 64
van Vliet, Erwin, 71, 77
Victoria (queen of England), 11
Visible Human Project, 76
vitamin A, 36
vitamin deficiency diseases, 35–36

White, Caroline Earle, 16
Why Animals Matter (Williams and
  Demello), 66–67
Williams, Erin E., 66
Wired (magazine), 29
Wolf, Gary, 29

xenotransplantation, 36–39, 78
  barriers to, 39–40

Ynterian, Pedro A., 53

Picture Credits

Cover: iStockphoto.com

AP Images: 65

Biophoto Associates/Science Photo Library: 37

Jean-Loup Charmet/Science Photo Library: 13

Klaus Guldbrandsen/Science Photo Library: 30

© Last Refuge/Robert Harding World Imagery/Corbis: 42

LTH NHS Trust/Science Photo Library: 74

Lynn McLaren/Science Photo Library: 50

Peter Menzel/Science Photo Library: 7

Hank Morgan/Science Photo Library: 25, 69

Ria Novosti/Science Photo Library: 21

Sinclair Stammers/Science Photo Library: 56

About the Authors

Bonnie Szumski has been an editor and author of nonfiction books for over 25 years. Jill Karson has been a writer and editor of nonfiction books for young adults for 15 years.